Sociocultural Evolution

SOCIOCULTURAL EVOLUTION

CALCULATION AND CONTINGENCY

Bruce G. Trigger

McGill University

First published 1998

2 4 6 8 10 9 7 5 3 1

Blackwell Publishers Ltd
108 Cowley Road
Oxford OX4 1JF
UK

Blackwell Publishers Inc.
350 Main Street
Malden, Massachusetts 02148
USA

Library of Congress Cataloging-in-Publication Data

Trigger, Bruce G.
    Sociocultural evolution: calculation and contingency / Bruce G. Trigger.
       p.   cm. — (New perspectives on the past)
    Includes bibliographical references and index.
    ISBN 1–55786–976–6 — ISBN 1–55786–977–4
    1. Social evolution. 2. Human evolution. 3. Social change. 4. Social history.
I. Title. II. Series: New perspectives on the past (Basil Blackwell Publisher)
HM 106. T75 1998
303.4 — dc21   97–29601
                                                      CIP

Typeset in Plantin 11 on 13 by York House Typographic Ltd
Printed in Great Britain by M.P.G. Books Limited, Bodmin, Cornwall

This book is printed on acid-free paper

Contents

Series Editor's Preface

History is one of many fields of knowledge. Like other fields it has two elements: boundaries and contents. The boundaries of history first acquired their modern shape in early modern Europe. They include, among other things, such basic principles as the assumption that time is divisible into past, present, and future; that the past can be known by means of records and remainders surviving to the present; that culture can be distinguished from nature; that anachronism can be avoided; that subjects are different from objects; that human beings are capable of taking action; and that action is shaped by circumstance. Above all else, of course, they include the assumption that history does actually constitute a separate field of knowledge that is in fact divided from neighbouring fields – not merely a hitherto neglected corner of some other field whose rightful owners ought ideally, and are expected eventually, to reclaim it from the squatters now dwelling there without authorization and cultivate it properly with the tools of, say, an improved theology or a more subtle natural science.

A prodigious harvest has been gathered from the field bounded by those assumptions. Making a tentative beginning with the humanist discovery of antiquity, gaining confidence with the Enlightenment critique of religion, and blossoming into full professionalization in the nineteenth century, modern historians have managed to turn their produce into an elementary ingredient in democratic education and a staple of cultural consumption. They have extracted mountains of evidence from archives and turned it into books whose truth can be assayed by anyone who cares to follow their instructions. They have dismantled ancient legends that had been handed down through the ages and laid them to rest in modern libraries. They have emancipated the study of the past from prophecy, apocalypticism, and other providential explications of the future. Pronouncements on the past no longer command respect unless they have been authenticated by reference to documents. Myths and superstitions have given way to knowledge of unprecedented depth, precision, and extent. Compared with what we read in older books, the books of history today are veritable miracles of comprehension, exactitude, and impartiality.

Success, however, has its price. None of the assumptions defining the modern practices of history are self-evidently true. The more they are obeyed, the less it seems they can be trusted. Having probed the realm of culture to its frontiers, we cannot find the boundary by which it is supposed to be divided from the empire of nature. Having raised our standards of objectivity to glorious heights, we are afflicted with vertiginous attacks of relativity. Having mined the archives to rock bottom, we find that the ores turn out to yield no meaning without amalgamation. And having religiously observed the boundary between the present and the past, we find that the past does not live in the records but in our own imagination. The boundaries of history have been worn down; the field is lying open to erosion.

The books in this series are meant to point a way out of that predicament. The authors come from different disciplines, all of them specialists in one subject or another. They do not proceed alike. Some deal with subjects straddling familiar boundaries – chronological, geographical, and conceptual. Some focus on the boundaries themselves. Some bring new subjects into view. Some view old subjects from a new perspective. But all of them share a concern that our present understanding of history needs to be reconfigured if it is not to turn into a mere product of the past that it is seeking to explain. They are convinced that the past does have a meaning for the present that transcends the interests of specialists. And they are determined to keep that meaning within reach by writing good short books for non-specialists and specialists alike.

Constantin Fasolt
University of Chicago

Preface

If anyone had told me thirty years ago that in the late 1990s I would be publishing a defence of sociocultural evolution, I would have laughed; first at the idea that a concept that was then so well entrenched would have required any defence, and second at the notion that I would be performing this task. But times change, even if individuals do not. In the 1960s I never imagined that the 1990s would be a time when highly productive Western economies would be accompanied by growing unemployment, lengthening bread lines, atrophying educational systems, lessening public care for the sick, the aged, and the handicapped, and growing despondency and disorientation – all of which would be accepted in the name of a nineteenth-century approach to economics that had been demonstrated to be dysfunctional already by the 1920s. These happenings alone might cause a person to despair of evolution.

Yet this is not the reason that sociocultural evolution is now under more severe attack than at any time in its history. I am surprised to find colleagues who I know to be intelligent and to have well-developed social consciences, especially with regard to third-world and indigenous peoples, rejecting the concept of sociocultural evolution in order to embrace an extreme cultural relativism that makes it impossible to evaluate the respective merits of any social practice. This way of thinking unwittingly encourages people to accept what we have as the best that is possible and to relinquish any hope of creating a better future. It is equally interesting to find a hegemonous neo-conservative economics totally uninterested in the concept of evolution and willing to propagate, if not to endorse, cultural relativism. There is a need to investigate the origin of such ideas, which is what I propose to do in this historical survey and evaluation of the concept of sociocultural evolution.

I know that by publishing this book I risk giving pain and moral offence to many people, especially my Wendat clan-brother Georges Sioui, who in his own writings has eloquently exposed the racism and colonialism inherent in so much evolutionary theorizing, and powerfully defended the relevance of the traditional beliefs of his people for the modern world (Sioui, 1992). I do not disagree with anything Georges has written on this topic and certainly have no wish to deny the negative role that sociocultural evolutionary theory has often played in the past. Yet the skill with which he has adapted his people's cultural traditions to the needs of the modern world provides an example of evolutionary development of a very different kind. Evolution is not, as some of its critics have maintained, the most dangerous idea in the modern world. On the contrary, there is at this stage in human history a strong need to believe in the possibility of being able to create a future that is better than the present. I maintain that a judicious view of sociocultural evolution can avoid the pitfalls of past conceptions and play such a role. Indeed, despite claims to the contrary, the concept of sociocultural evolution has been progressively refined since the eighteenth century and is now better suited than ever before to play such a role.

I would also note that, despite the claims of certain critics (e.g. Renfrew, 1988: 198–200), I have never attempted to explain the development of evolutionary or other concepts entirely in terms of social context. While I regard social context as important, much of this book is concerned with the development of sociocultural evolution as an intellectual process; hence it is also examined from an internalist point of view.

Part of the research for chapters 2 to 8 was done in connection with writing *A History of Archaeological Thought* (Trigger, 1989). The themes of these two books are, however, very different, and, while some of the same material appears in both, it is here employed in a different manner and alongside new data.

I do not seek to provide comprehensive coverage of the literature dealing with sociocultural evolution. More has been written on this topic than could be digested in a lifetime. Hence many important contributions will go unmentioned. I also realize that a short work such as this cannot do justice to all the complex debates relating to sociocultural evolution that have occurred over the years. It is, however, my aim to avoid a presentistic or misleadingly unilinear account. The most that is possible is to give an impression of the vast scope and complexity of this topic. More detailed accounts are available elsewhere (Bowler, 1989; Burrow, 1966; Sanderson, 1990).

While I acknowledge that there have been close connections and much intellectual cross-fertilization between biological and socio-cultural evolution, my coverage of these ties will be limited to what I judge is necessary to understand sociocultural evolution. This relationship, like everything relating to sociocultural evolution, has been complex and controversial.

Finally, I will assess sociocultural evolutionary thought more in relation to archaeology than to the more familiar reference points of economics, political science, sociology, and anthropology. This is partly because archaeology is the discipline I know best, but also because archaeological findings are extremely useful for evaluating evidence relating to long sequences of sociocultural change. These archaeological examples remain a means to an end, which is to evaluate sociocultural evolutionary theory, the role it has played in Western thought, and its relevance for coping with current and future problems.

I wish to acknowledge the intellectual stimulation I have received from Ursula Franklin, especially her wonderful book *The Real World of Technology* (Franklin, 1990), and from reading, as I was doing in the final stages of writing this book, my colleague Gregory Baum's (1996) *Karl Polanyi on Ethics and Economics*, a far more widely ranging essay than its title suggests. I am also aware, now that it is nearly finished, of the extent to which this book is part of a Canadian social science tradition which refuses to separate scholarship from social and moral concerns. The great exemplar of this tradition was the economic historian Harold Innis.

I also wish to thank Constantin Fasolt for drawing my attention to the work of Norbert Elias and for his extremely stimulating comments on my work; my friend Alan H. Adamson, Professor Emeritus of History at Concordia University, and Stephen Chrisomalis, a doctoral student in anthropology at McGill University, for commenting on a draft of this book; and Dr Barbara Welch both for critiquing the manuscript and for an unparalleled chance to share and discuss ideas. I am grateful to John Davey for persuading me to write this book and to Tessa Harvey for seeing it through the press.

This book was finished while I was the recipient of a sabbatical leave from McGill University. It is dedicated to the memory of my grandfather, Harry Graham, of Preston, Ontario. A latter-day son of the Scottish Enlightenment, he was guided by his faith in the essential goodness of human beings, the reality of social progress, and the fundamental importance of personal responsibility. His life bears witness to the power of these beliefs to inspire an individual to strive for the common good.

1
Introduction

the intellectual is an individual endowed with a faculty for representing,
embodying, articulating a message, a view, an attitude, philosophy or opinion
to, as well as for, a public. And this role has an edge to it, and cannot be played
without a sense of being someone whose place it is publicly to raise embarrassing
questions, to confront orthodoxy
Edward Said, *Representations of the Intellectual* (1994) p. 11

For over two hundred years claims made for and against sociocultural evolution have influenced major debates about human
behaviour and social policy in Western civilization. Long before
biology became a bastion of evolutionary thought, evolutionary
concepts played a key role in discussions of social, political, and
economic issues. The idea that progressive change, rather than
stability, decay, or endless cycles, was the essence of the human
condition was a revolutionary concept that called into question
traditional European beliefs and undermined many of the constraints that had guided European life prior to the eighteenth
century. The concept of sociocultural evolution was born amidst
political controversy and has remained controversial, although the
emphasis of these debates has shifted over time.

Sociocultural evolution can be characterized as a way of looking
at human history. It implies an overall shape or direction to history
that can be explained rationally and which endows human behaviour with special meaning. It also assumes that at least one aspect
of directionality is a trend towards greater complexity. The idea of
sociocultural evolution is considerably older than its name, which
is derived from the Latin *evolutio*, referring to the unrolling of a
scroll or parchment. In the eighteenth century, what is now called

sociocultural evolution was generally referred to as the doctrine of progress or progressive transformation. In the 1850s, Herbert Spencer adopted the term 'evolution' as a synonym for progress. Hitherto, naturalists such as Charles Lyell had begun to use it more narrowly to designate the then highly controversial notion of the origin of biological species by development from earlier forms (Ingold, 1986: 4; Sanderson, 1990: 28). Spencer maintained that progressive change characterized all things and was equally applicable to the physical, biological, and psychosocial realms. Harriet Martineau, the British advocate of unregulated free enterprise, referred specifically to social evolution as early as 1853, while Spencer popularized her designation almost a decade later.

Although initially progress and evolution were treated as synonymous, following the spread of Darwinian theory, biological evolution came to signify all forms of organic descent with modification, not simply ones that resulted in greater complexity. It took longer for evolution to acquire this broader meaning in the social sciences, and in these disciplines the older meaning tended to survive alongside the newer one. Despite changing definitions, the main aspect of sociocultural evolution that continues to interest both the general public and social scientists remains that of progress. People want to know whether or not humanity as a whole is progressing. Or perhaps, more to the point, in what ways is humanity progressing or not progressing? While these questions will be shown to be misphrased, they address an issue that is of more than purely subjective concern.

A Controversial Doctrine

The concept of progressive change was not new in the eighteenth century. As early as the sixth century BC, the Greek philosopher Heracleitus recognized the universality of change when he stated that it was impossible for a person to step into the same river twice. This broke with an older tradition that there was nothing new under the sun. Still earlier Anaximander of Miletus had proposed that humans had been produced by an oviparous species of whale. In his long poem *De Rerum Natura* (On the Nature of Things), written in the first century BC, the Roman Epicurean philosopher Titus Lucretius Carus outlined a comprehensive theory of evolution in which primordial atoms spontaneously combined to create ever more complex physical and then biological entities. He depicted human society as evolving through successive ages of stone, bronze, and iron. Yet in classical civilization such evolutionary ideas remained speculations, alongside alternative degenerationist, static, and cyclical views of human history. For example, the Roman emperor Marcus Aurelius believed that, if a man lived to be forty, he would have witnessed all that had happened in the past or could hope to be seen in the future (Roberts, 1996: 275). All the different pasts that could possibly be imagined were conceptualized, at least in a general fashion, at this time, but no one appears to have given any thought to finding ways to determine which of these alternatives corresponded with reality. Even the historically documented change from a bronze- to an iron-based technology was interpreted both as an evolutionary advance and as evidence of social and moral degeneration from an original Golden Age. In the later Roman Empire the concept of eternal Platonic essences counted for much more than did the idea of cultural or biological evolution.

It was only in the context of accelerating technological, economic, and cultural changes in Western Europe during the eighteenth century that the idea of progressive sociocultural change became something more than a subject for idle speculation. A growing number of intellectuals, reflecting on the developments of the preceding centuries, began to believe that what was new might be better, not worse, than what had gone before, and gained confidence in the ability of human beings to create a better way of life for themselves. The doctrine of progress became a weapon for demanding the abolition of absolutism and the surviving privileges of the feudal order, which a growing number of the professional and commercial middle class saw as impediments to their own interests. Sociocultural evolution provided the intellectual basis for justifying demands for social reforms and, where, as in France, these demands were not met, it could also be used to justify revolution.

Because of its political implications, sociocultural evolution was attacked almost immediately as a flawed, inadequate, and socially pernicious doctrine. The attacks came not only from supporters of traditional regimes but also from romantics who continued to regard the past as providing the ideal model for the present and who believed that evolutionary concepts trivialized cultural diversity and hence those traditions and associations that provided human life with much of its meaning. Since changes do not benefit everyone equally and all people do not share the same values and opinions, it is not surprising that people fail to agree whether they live in the best or worst of times. Yet, by challenging people to consider the general direction in which societies are moving, the concept of sociocultural evolution has subjected entrenched self-interest and cultural prejudices to new and often unwelcome scrutiny.

Today, when the world is experiencing unprecedented technological and social change, evolution is being reviled more vociferously than ever before as a dangerous and scientifically flawed concept. Those who are alarmed or morally outraged by what is happening denounce evolution for celebrating progress in a world in which unchecked technological development and economic exploitation are producing human carnage and ecological destruction on an ever-increasing scale. Many people profess that the present is not the most civilized but the cruellest, and hence most degenerate, phase of human existence. Evolution is alleged to be an inherently ethnocentric, and often racist, legacy of imperialism, which denigrates non-Western peoples by equating their modern cultures with the primitive past of Western civilization.

In their violent opposition to such alleged denigration the authors of *Maya Cosmos* assert that 'what our nineteenth-century European and American ancestors took for global destiny – progress and enlightenment of the world through acquaintance with the Western perception of culture – was a historical happenstance. As the domination of European and North American society slowly dissipates in the twilight of the twentieth century, those still-remaining, vibrant cultures throughout the world are now reaching out to share the technological and economic power we thought was ours alone' (Freidel et al., 1993: 36). Thus sociocultural evolution is identified not as something real and continuous but as a Western intellectual construction of the colonial era that is no different from any other 'tribal belief'. It is also maintained that evolutionary approaches erroneously confuse change with progress, ignore the roles played by external influences and human agency, project inappropriate modern concepts into the past, and seriously distort the understanding of processes of social and cultural change. They are also alleged to bias narratives of the past in favour of pro-capitalist or pro-Marxist options, thus forcing them to follow hegemonous Western thought patterns (Johnson, 1996: 68). These criticisms are intensified by a commitment to postmodern subjectivism, which stresses the idiosyncratic and disjunctive aspects of understanding and the degree to which beliefs are consciously or unconsciously shaped by self-interest.

These critics of evolution, like the romantics before them, are reacting defensively and indiscriminately to what they perceive as the threat that sociocultural change poses to human values. They do this by resorting to nostalgic introspection and retrospection, supporting collective values, and emphasizing the desirability of cultural diversity. At a deeper level, they seek to undermine the claims of modernity by espousing an extreme relativism that denies it is possible to judge one way of life, or any particular idea, as being better or worse than any other. Relativists argue that every culture must be understood on its own terms and that comparisons between them are bound to be rendered inappropriate by the ethnocentric prejudices of the analyst. The price that must be paid for adopting this position is to make it impossible to condemn and suggest alternatives for even the most selfish and socially destructive ideas and behaviour, except on purely subjective grounds. It leaves no basis on which to act for the betterment of all human beings. Postmodern relativism is not only defensive but also defeatist.

On the other hand, one might have expected more enthusiasm for evolutionary thought among neo-conservatives. The 1980s and 1990s have witnessed the apparently definitive triumph of capitalism over Marxism and the unprecedented expansion of free-enterprise economic activity everywhere in the world. Surely these heady developments ought to have expressed themselves in a renewed celebration of progress and of the vast transformation that capitalism itself has been undergoing as a result of the increasing importance and power of transnational corporations. Yet this has not happened. Capitalism has not developed a new philosophy and ideology but, under the label of neo-conservatism, has succeeded only in reviving the long-discredited doctrines of nineteenth-century liberalism. The ideologues of the New Right are more intent on emphasizing what they view as the essential and unchanging aspects of capitalism than on acknowledging the major changes that are transforming both it and the modern world. Capitalism is being portrayed as an 'end of history' (Fukuyama, 1992), a form of economic organization whose essential features are inherently perfect and hence unchanging.

It might be imagined that this lack of interest in evolution is simply the result of a myopic formulation that promises short-term advantage for some individuals who act in their own self-interest, while leaving it to an invisible hand and an ill-defined trickle-down effect to promote the general good. Yet the Victorian liberals who held these same beliefs were enthusiastic exponents of socio-cultural evolution. It seems more likely that the silence of neo-conservatives about general patterns of change reflects their own doubts about the ability of a free-enterprise economy to produce widespread and lasting prosperity on a global scale. The revival of long-discredited economic theories to justify the rapacity of individuals and corporations indicates a lack of creativity in devising ways to overcome the social disruption that capitalism has produced in the past, whenever individuals and corporations were able to pursue their own goals free of constraints imposed on them by nation states in an effort to protect the public good.

Thus sociocultural evolution is being rejected for apparently different reasons by neo-conservatives and their relativist critics. Those who oppose neo-conservatism tend to regard technological and economic change as inherently dangerous and seek to protect traditional lifestyles by defending cultural variation and collective values. Neo-conservatives try to disguise their own realization, based on past experience, that modern, politically-disembedded, transnational capitalism is even more destructive to the public good than were earlier forms by emphasizing the most generic characteristics of free enterprise rather than treating it as a system that is constantly changing. This also allows them to avoid discussing the question of whether a transnational economy might eventually evolve into a form of economic organization that would no longer display all the essential features of capitalism.

What postmodern romantics and neo-conservatives share in common, and leads both to abandon sociocultural evolution, is their lack of confidence in the future, and even their fear of it. Each approach in its own way damages prospects for the future because it discourages research that might lead to a more detailed and realistic understanding of sociocultural change and a more informed public discussion of the range of choices that are open to human societies, and what the consequences of each option might be. Despite their seemingly irreconcilable ideological differences, there is an unadmitted and largely unconscious affinity between neo-conservatism and postmodernism, which has led to postmodernism being described as the 'cultural logic of late capitalism' (Harvey, 1990: 42). Postmodernism's most striking characteristic is its encouragement of an iconic intellectual environment which destroys a sense of time as unfolding connections and hence promotes a view of the past as being nothing more than 'unsubstantiated rumour' and of the future as an 'unacknowledged responsibility' (Solway, 1997). The key to the affinity between postmodernism and neo-conservatism is that both movements have become to a considerable extent embodiments of romanticism. While postmodernism is more or less the direct heir of earlier forms of romanticism, neo-conservatism, despite its claims to be modelled on rationalist philosophy, has grown increasingly more romantic as a consequence of its doubts about its ability to achieve its professed objectives.

The real antithesis of evolutionary studies is not a relativist celebration of cultural differences but a reluctance to study important issues that relate to human well-being. While neo-conservatives exaggerate the essential and unchanging features of capitalism, extreme relativists discourage the study of problems on more than the local level for fear that more extensive and comparative studies might be construed as racist or colonial. Yet ruling out the study of such questions inhibits social scientists from gaining essential insights into the larger economic, social, and political structures that underlie the hegemony of modern capitalism (Sherratt, 1993: 125). Thus, ironically, extreme relativism, far from opposing neo-conservatism, helps to blunt a critical study of its operation and the creation of effective alternatives to it. Far from sociocultural evolution being a doctrine primarily associated with oppression and exploitation, as its critics claim, for over 250 years it has been promoted mainly by those who have sought to encourage progress and social change, while romantically based cultural relativism and essentialism have been the doctrines of choice for those who fear and oppose progressive social change.

Definitions of Evolution

I use the cumbersome term sociocultural evolution because the processes that shape the social networks in which all human beings live are different from those which alter the cognitive patterns underlying human behaviour (Flannery, 1972, 1995). At the same time, because all human interaction and behaviour are cognitively mediated, their social and cultural aspects cannot be studied in isolation from one another.

Although most people understand in a general sense what is meant by sociocultural evolution, there are major disagreements concerning its precise conceptualization. Opinion divides as to whether evolution refers to actual processes of social and cultural change, to the mechanisms that bring about such change, or to both. George P. Murdock (1959) and V. Gordon Childe (1947), as well as most classical Marxists, have equated evolution with the totality of what has happened in the course of human history, thus rendering the subject matter of history and evolution identical. This is analogous to biologists using the term evolution to refer to the total pattern by which all species that have ever lived are related to one another. Murdock considered efforts to delineate and explain patterns of continuity and change within the totality of evolution to be exercises in deductive philosophy.

Others equate sociocultural evolution not with the totality of human history but only with specifically linear and progressive patterns that can be detected within that totality. Marvin Harris (1968) equates evolution with any structural transformation of a sociocultural system. Yet, while he pays little attention to the trends that have fascinated unilinear evolutionists, he argues that parallel and convergent development are more common in socio-cultural evolution than is divergent development (Sanderson, 1990: 154–5). Most evolutionists equate sociocultural evolution with advances towards higher levels of organization, which normally means towards societies that are larger in scale and exhibit greater internal differentiation (Bowler, 1984: 9). Serious problems are encountered in reconciling this approach, which tends to be concerned primarily with defining the various stages through which human societies may evolve, with the cultural variation that results from adaptations to different environments. In view of the importance that biologists attach to adaptation as a force shaping biological evolution, it is difficult for social scientists to ignore adaptation as a factor influencing sociocultural evolution.

Still others equate sociocultural evolution neither with the totality of human history nor with specific patterns of change but with the mechanisms that bring about such change. This equates evolutionary theory with explaining how social and cultural change occurs in societies at all levels of development. Robert Boyd and Peter Richerson (1985: 289) specifically correlate it with accounting for how cultural variation is transmitted, expressed by individual human beings, and modified with descent from one generation to another.

In biology, however, the basic mechanisms of evolution have been confirmed and elaborated over the past 150 years. Biologists continue to debate issues such as group selection, the relative importance of gradual change and punctuated equilibrium, and the role played by catastrophes of extraterrestrial origin, but in general they agree concerning what brings change about. In the social sciences there is no generally accepted single explanation of change; instead there is a spectrum of competing theories which stress ecological determinism at one end and cultural determinism at the other. It is not even agreed whether the mechanisms of change remain constant, or change as societies become more complex. While ecological determinists tend to view the same factors (environment, technology, and population) as accounting for the general nature of change in societies at all levels of complexity, Marxists argue that the process of change alters significantly as societies become more complex, thereby introducing what Gordon Childe (1947) has called real novelty into the course of human history. Stephen Sanderson (1990: 165–6) likewise argues that with the rise of capitalism the determining factors shifted from demographic and ecological to political-economic ones.

It is reasonable to maintain that the study of sociocultural evolution must include explanations of change, and it may make sense to equate evolution primarily with such mechanisms. Yet it is impossible to construct theories of change independently of what they purport to explain. There would be no point to trying to formulate evolutionary theories if it were clear that all cultural change took place in either a cyclical or a purely random fashion. Evolution is concerned primarily with understanding directionality as a major characteristic of human history. This directionality involves an overall tendency towards creating larger, more internally differentiated, and more complexly articulated structures that require greater *per capita* expenditures of energy for their operation. Hence this complexity increases in defiance of the second law of thermodynamics (Adams, 1988). A concern with increasing complexity does not, however, imply that sociocultural evolution is continuous, universal, irreversible, or the result of anything inherent either in human nature or in the structure of the cosmos. Nor does it mean that change is bound to take place everywhere at a steady rate or even at all. All of these ideas have been advocated by sociocultural evolutionists at one time or another, but none has won lasting support. Moreover, no evolutionary theory maintains that sociocultural complexity is likely to go on increasing forever, any more than it is believed that terrestrial life or the cosmos can endure endlessly. Nor, despite claims to the contrary (Mann, 1986), does an evolutionary approach's concern with cross-cultural uniformities exclude a singularity, such as the emergence of capitalism in Western Europe, from being explained using evolutionary principles. Where specific developments have sufficient impact to rule out the possibility of similar developments occurring elsewhere, multiple instances may not be forthcoming. Singularity cannot exclude changes which may transform the entire world system from being treated as examples of an evolutionary process.

Despite these disputes, the term 'sociocultural evolution' tends to be used colloquially to refer both to the course of human history (in either its overall complexity or its general tendencies) and to the mechanisms that shape that history. This dual usage corresponds to the way the concept of evolution is used by biologists. Evolution thus refers to both a mechanism and its output. More specific definitions become important only in so far as they influence the scientific study of change and variation. In the case of sociocultural evolution the main debate concerns whether or not there is an overall pattern to human history that requires explanation in addition to accounting for the processes that bring about social and cultural change on a daily basis.

Something else that is not agreed is the extent of regularity in human behaviour and sociocultural change. Extreme evolutionists have traditionally stressed the prevalence of cross-cultural regularities at the expense of cultural idiosyncrasies. Some evolutionists, such as Julian Steward (1955: 209), have tried to stack the deck in favour of such an approach by claiming that a science of human behaviour need consider and seek to explain only the cross-cultural regularities in human behaviour, leaving the 'unique, exotic, and non-recurrent particulars' to be studied by culture historians. Anti-evolutionists have been equally adept at ignoring regularities. By arguing that all human behaviour is cognitively mediated and can be understood only in relation to the totality of the culture of which it is a part, they often seek to deny that it is possible to study human behaviour cross-culturally.

Anthropologists have long been caught on the horns of a

dilemma of their own making. On the one hand, their conviction that all human groups are intellectually and emotionally similar encourages support for an evolutionary position. On the other hand, their advocacy of cultural relativism leads them to celebrate cultural diversity and to argue that every culture is the product of its own history, and influences the behaviour of its members in a unique fashion. As a result many anthropologists oppose cultural evolution's search for cross-cultural uniformities. With rare exceptions, partisanship has inhibited detailed comparisons of how similar or different cultures actually are. More empirical work along these lines must be done, especially with respect to cultures that appear to be at an equivalent level of development, before it is possible to make evidentially grounded statements about how cross-culturally similar or specifically different cultures are in respect of particular features (Trigger, 1993).

This book has two closely related goals. The first is to survey the study of sociocultural evolution, together with the social and political uses that have been made of this concept from the eighteenth century to the present. The second is to demonstrate that sociocultural evolution is a valid, and indeed an essential, concept that can contribute to the better understanding and management of human affairs. A historical survey is especially important because of the tendency for new social theories to build on partial and distorted versions, often to the point of caricature, of the concepts they are trying to replace. This is as true of new evolutionary theories attempting to replace old ones as it is of evolutionary and anti-evolutionary theories trying to discredit each other. A selective emphasis on the weaker features of older theories results in significant aspects of older social science thought being discarded even when they remain valid and useful. It also results in an inappropriate emphasis on the discontinuities and weaknesses that have characterized sociocultural evolutionary approaches over the past 300 years and minimizes the progress that has been made. This study will demonstrate that there has been both more continuity and more progress in developing an understanding of sociocultural evolution than is generally thought to be the case. Significant developments include the widespread abandonment of

the concept of progress on the grounds that it is inherently subjective and value-laden; the increasing integration of concepts of
adaptation into sociocultural evolution; and the abandonment of
transcendental teleologies.

In spite of this, it is clear that subjective factors have strongly
influenced the study of sociocultural evolution and continue to do
so. In general, the most enthusiastic support for evolutionary
theory has occurred when the power of the middle class was
increasing or unchallenged in different parts of the world. Periods
of political crisis for the middle class have been characterized by
the rejection of the concept of sociocultural evolution or by its
construal as an inimical force. The following chapters will document the various cycles through which the concept of sociocultural
evolution has passed since the eighteenth century. The reason for
this parallelism between sociocultural theorizing and the economic
and political fortunes of the middle class is not hard to discover.
While the members of the intellectual middle class who have
propounded both theories of sociocultural evolution and their
antitheses do not necessarily share precisely the same point of view
as do members of the industrial, commercial, or professional
middle class, their thinking reflects the fortunes of the middle class
in general, which have altered rapidly over the past 300 years as a
result of fluctuations within an ever-changing capitalist economy
(Marx [1869] in Marx and Engels, 1962, 1: 275).

Our historical survey suggests, however, that, despite a strongly
subjective element in the interpretation of data relating to human
behaviour and a tendency for interpretations to be formulated,
either consciously or unconsciously, in order to promote specific
social and economic interests, interpretations of human behaviour
and human history are gradually becoming more comprehensive
and objective. This is because evidence, which exists independently of the observer, has the capacity to resist, however feebly and
erratically, both falsifications and innocent misinterpretations
(Trigger, 1989; Wylie, 1982, 1985, 1993). Despite the current
ascendancy of both relativism and neo-conservative ideological
bias, and the impossibility of knowing to what extent any specific
interpretation corresponds with reality, this increasing knowledge,

which is itself part of the trajectory of sociocultural evolution, is becoming increasingly important for formulating public policy as growing human populations and a worldwide desire for materially enhanced living standards press ever harder on the finite resources of the planet. It is neither necessary nor desirable to submit to the intellectual paralysis to which extreme relativists and right-wing essentialists would reduce the social sciences.

2
Reversing Utopia

The point of departure for investigating the history of sociocultural evolution is the view of history that prevailed in medieval Europe. While, as we have already noted, the concept of sociocultural evolution existed in classical times, the medieval view of history constitutes the background against which modern evolutionism emerged and defined itself. Although many of the ideas associated with the medieval view of history remain part of Western civilization, the paradigm itself and the intellectual assumptions that sustained it are sufficiently different from those of modern Western civilization that understanding them requires no less an act of anthropological imagination than does gaining insight into the cosmologies of non-Western peoples.

Medieval Cosmology

Contrary to widespread belief, the medieval period in Western and Central Europe (AD 476–1500) was not so much a time of ignorance and stagnation as one of extreme political fragmentation. Supporting scholarship was the responsibility neither of feudal rulers, who were hard pressed to keep order among themselves, nor of the craftsmen and traders who lived in the small urban centres. Learning was carried on in the monasteries and universities founded by the Roman Catholic Church, which as part of its mission worked to maintain civilization as well as a sense of community across a politically disunited continent. Theology, philosophy, and law were the principal preoccupations of medieval scholarship, although efforts were also made to conserve the practical scientific knowledge, and in particular the medical knowledge, that was the heritage of classical civilization.

Intellectual life was subject to the control of the Christian Church, which held strict views about what individuals must believe in order to assure their personal salvation. This attitude eliminated, at least from public discussion, much of the free-wheeling speculation that had characterized intellectual life during the classical period. The scriptures, the teachings of the Church, and at least such surviving classical thought as appeared not to contradict Christian teaching constituted the basis of medieval scholarly thinking about the closely linked topics of cosmology and human history.

The earth was believed to be located at the centre of the cosmos and to be small, recently created, and destined not to last for more than a few thousand years. Most Europeans imagined the world to be flat, with a sky and supernatural heaven, presided over by God, above, and a fiery place of punishment inhabited by devils below. Scholars who were familiar with classical learning understood that the earth was spherical and believed it to be embedded in a series of revolving crystalline spheres that were associated with various heavenly bodies. On the basis of scripture, it was thought that the universe had been supernaturally created over the course of six days by God speaking commands that had transformed chaos into the existing order of things. Rabbinical authorities estimated from the genealogical data contained in the Old Testament that the universe had been created around 3700 BC, although Pope Clement VIII (r. 1592–1605) later dated its beginning to 5199 BC. The idea of a god creating the universe by issuing verbal commands long antedated both Judaism and Christianity. It reflected a view of how things got done on a grand scale in kingdoms whose rulers issued commands to their subordinates.

It is often maintained that Christian ideas played a major role in the development of an interest in history in Western civilization.

The Bible offered a detailed, generation by generation narrative of what was supposed to have happened in the Middle East from the time of the Creation into the early Christian era, and endowed these events with cosmic significance. Christian thought therefore favoured a linear over a cyclical view of time. But, within the context of Christian religious history, the only freedom assigned to individual human beings was the power to choose between good and evil; a choice that determined the everlasting destiny of their souls after death. The course of human history was fixed and major transformations took the form of a series of unique events that came about as the result of divine intervention. Creation was followed by humanity's fall into disobedience to God and then, after several thousand years, by Christ's incarnation and death, which saved human beings from their original sin. This in turn was to be followed by the return of Christ and the Last Judgement, the destruction of the present cosmos, and its replacement by a perfect and unchanging spiritual order. Major turning points in human history, apart from original sin, were the result of divine intervention, and human history represented the unfolding of a plan that God had in mind before he created the universe. There was no awareness that any form of positive directional change might be intrinsic in human history or that human beings, except when acting as God's agents, could achieve anything of historical significance.

There was no more idea of a natural order than there was of an order that was intrinsic to human history. God made, but could also abrogate or alter, the laws of nature, and frequently did so by performing miracles. Therefore, neither human beings nor the world in which they lived had an order that was not subject to divine transformation in both major and minor ways. Human and cosmic history were by-products of God's plans for humanity's individual and collective spiritual destiny.

Medieval Christians believed that Christ's second coming was likely to occur within a few hundred years. Hence the universe was thought not to have been intended to last more than a total of seven or eight thousand years and humans believed themselves to be living near its last days. In accordance with this view, it was generally accepted that the resources of the world were being quickly depleted. Mines were being exhausted, water and wind erosion was removing fertile soil, and even carefully tended agricultural land produced smaller harvests as time passed. It was widely believed that the physical world was in an advanced state of decay (Slotkin, 1965: 37; Toulmin and Goodfield, 1966: 75–6). Given that God had not intended the world to last for more than a few thousand years, this depletion of its resources did not seem strange. It was in effect merely another manifestation of the transient and corrupt nature, and spiritual insignificance, of all the things of this world.

The biblical documentation of greater human longevity in antiquity, when human beings were claimed regularly to have lived many hundreds of years, and its mention of races of giants suggested that humans had been physically deteriorating since Creation. It was also accepted that the impossibility of duplicating classical works of scholarship, art, and architecture indicated that the human intellect had weakened as well. Medieval scholars viewed themselves as living in the twilight of the world and human intelligence as diminishing along with the rest of God's Creation. Without God's revelations and the Church to provide guidance, society and morals also would deteriorate rapidly. This was a local variant of the widely shared belief in early civilizations that the gods had created a perfect society at the beginning of time, which it was the duty of later generations of humans to try to emulate. In the view of medieval Christians, neither the individual nor humanity as a whole could escape damnation without God's active intervention on their behalf.

Human beings were believed to have been created by God in the Garden of Eden, which was located somewhere in the Middle East. From there they had spread to other parts of the world, first after the expulsion of Adam and Eve from Eden, which they had tended for God, and again following the destruction of all but a handful of humans by Noah's flood. On the basis of biblical accounts, it was accepted that from the time of their expulsion from the Garden of Eden human beings had grown crops and practised animal husbandry and that soon afterwards they were working bronze and

iron and living in cities. Their second dispersal was hastened by
God's differentiation of languages as retribution for human pre-
sumption in trying to build a tower that reached up to heaven. The
centre of world history remained in the Middle East until the
beginning of the Christian era.

The earliest humans shared in God's direct revelation of himself
to Adam, the first human, and knowledge of God and his wishes
was maintained and elaborated through divine revelations made to
successive Hebrew patriarchs and prophets. These, together with
the revelations contained in the New Testament, became the
property of the Christian Church, which henceforth was responsible for upholding standards of human conduct and ensuring
individual salvation. On the other hand, groups that had moved
away from the Middle East and failed to have their faith renewed
by divine revelation or Christian teaching tended to degenerate
into polytheism, idolatry, and immorality. Hence it was possible
for medieval scholars to view the pagan Romans of antiquity as
physically, intellectually, and culturally superior to themselves
because they had existed closer to the creation of the world, but
also as inferior because they had lost the knowledge of divine
revelation which Christianity had restored to the peoples of
Europe.

There was also little awareness of social and cultural change. A
few popes and emperors, such as Charlemagne and Frederick
Barbarossa, collected ancient gems and coins, reused elements of
Roman architecture, and occasionally imitated Roman sculpture
(Schnapp, 1993: 89–104; Weiss, 1969: 3–15). But, in general,
medieval artists assumed that in classical and biblical times people
wore clothes and lived in houses that resembled their own. Even
growing contacts with Islamic civilization during the crusades did
not suggest that in biblical times people might have been other
than socially and culturally similar to those of medieval Europe.

Yet medieval intellectual life was more complex than is commonly assumed. People were aware that they had improved, and
were continuing to improve, their technology. The twelfth-century
Italian theologian, Joachim of Fiori, divided human history into
three progressive stages, of which the third, or age of the Holy

Spirit, was to be reached by the Church only after great tribulation (Reeves, 1969). Nevertheless, so great was the scholarly emphasis on moral and spiritual issues that learned people paid little attention to the technological and economic progress that was slowly increasing productivity and enhancing the quality of life over large areas of Europe.

Accommodating Change

By the fourteenth century the increasing rate of economic growth and rapid social change were undermining feudal institutions in northern Italy and leading to the emergence of a series of city states. Scholars, whose views generally reflected the interests of the rising nobility and bourgeoisie on whose patronage they depended, attempted to justify these innovations by demonstrating that they had precedents in ancient times. In keeping with the medieval pattern of thought, far from praising what was happening as novel, they sought to demonstrate that the emerging Italian city states resembled those of classical Italy and Greece (Slotkin, 1965: x).

The use of historical precedent to justify innovation slowly led to the realization that contemporary social and cultural life did not resemble that of classical antiquity and that the past had to be understood on its own terms. As a result of growing familiarity with the historical, literary, philosophical, legal, and scientific texts of ancient Rome and Greece, many of which had been unknown and hence unstudied in Western Europe since the fall of the Roman Empire, scholars tried to understand classical civilization. The appreciation of classical antiquity also rapidly extended into the fields of art and architecture, where surviving structures and art works were more informative than literary texts (Jacks, 1993). The aim of Renaissance scholars was to understand and to try to emulate as best they could the glorious achievements of antiquity. For Petrarch and many later Renaissance scholars the period that separated classical civilization from their own age was an age of barbarism and cultural deprivation that was unworthy of study. This involved what is now regarded as a purely subjective devaluation of the engineering and artistic skills of medieval architects and artists. At first there was little faith that in their current degenerate state human beings could ever hope to equal, let alone surpass, the achievements of antiquity. Yet, under the impetus of classical studies, Gothic art and architecture gave way to classical revival styles and writers began to emulate, and even try to surpass, the great literary works of antiquity. Michelangelo Buonarroti boasted of his plans for the church of San Giovanni dei Fiorentini that 'neither Greeks nor Romans ever constructed such a thing among their temples' (Clements, 1974: 100). In the philosophical and religious realm, however, there was at this time little challenge to a belief in degeneration as a general characteristic of the physical world and of secular culture.

In the literature of ancient Greece and Rome Renaissance scholars encountered cultures that had no single view of antiquity and in which theories of cultural stasis, degeneration, cyclical change, and evolution competed on equal terms as explanations of the past. The aim of all but the least prudent of these scholars was to accommodate these views as far as possible to orthodox Christian ones.

The easiest of these views to accommodate was degeneration. Classical theories of a primordial golden age of leisure and innocence which lapsed into increasing violence and toil through successive ages of silver, bronze, and iron seemed to parallel in some metaphorical fashion the biblical account of Adam and Eve's expulsion from the Garden of Eden. Occasionally speculation about degeneration strayed beyond the bounds of Christian orthodoxy. Some scholars came to believe that the Greek hermetic texts, which first became known in Italy in the fifteenth century, contained ancient Egyptian wisdom that might antedate the Hebrew scriptures and hence offer an older, more complete, and more authentic account of divine revelation than did the Bible. While these views were tolerated by Italian Church authorities during much of the sixteenth century, they became unacceptable as a result of Counter-Reformation demands for greater theological orthodoxy, and were driven underground after their chief exponent Giordano Bruno was burned at the stake in Rome in 1600 (Iversen, 1993; Yates, 1964).

More orthodox degenerationists believed that knowledge of revealed religion and of technological skills such as agriculture were lost as humanity broke into small groups after the confusion of languages that followed the attempt to build the tower of Babel. According to the Bible, agriculture and animal husbandry were known to the first human beings and iron-working was recorded as practised soon after. Since the Church taught that all humans were descended from Adam and Eve, peoples who lacked such knowledge were assumed to have lost it. Yet another insight into degenerationism was provided by the rediscovery in the fifteenth century of the Roman writer Cornelius Tacitus' *Germania*, which contained a polemical comparison of the rustic but noble Germans with what its author portrayed as the corrupt and effete Romans of his own day. This study suggested that increasing prosperity and cultural refinement might be accompanied by moral decay.

The evolutionary views of antiquity were echoed in the Jesuit missionary José de Acosta's *Historia natural y moral de las Indias*, published in 1589. Noting that there were hunter-gatherers, primitive agriculturalists, and civilized peoples living in the New World at the time of European discovery and believing them all to have arrived there as wandering hunters coming across the Bering Strait, Acosta interpreted the more complex societies as having evolved from hunter-gatherer ones after they reached North America. He squared this view with conventional Church teachings by arguing that the first inhabitants of the New World must have lost all knowledge of agriculture and iron-working as they made their way across Siberia and Alaska (Pagden, 1982: 193–7).

Cyclical views also were inspired by classical models, although these became more popular in the eighteenth century. The art historian Johann Winckelmann in his *Geschichte der Kunst des Altertums* (*History of Ancient Art*) ([1764] 1968) described classical art as progressing from a stage of infancy to one of maturity in classical Greek times and then in Roman times falling into a stage of decadence. Earlier in the century the Italian philosopher Giambattista Vico had characterized human societies as passing through a uniform cycle of development and decay that he believed reflected the regular working of providence. This cycle consisted of barbarism followed in turn by political organization based on fear of the supernatural, paternalistic government, and egalitarian society. The latter led to growing corruption and most likely to a reversion to barbarism. Vico squared his position with religious orthodoxy by affirming that degeneration was possible only among groups that had emigrated from the Middle East and lost their knowledge of true religion. The sacred history of the Hebrews recounted in the Bible was exempt from this cycle and set a constant and divinely-guided model for all human behaviour (Rossi, 1984). Although Vico was not an evolutionist, his views encouraged the belief that human history could be understood in terms of regularities similar to those being proposed by the natural sciences.

Cyclical views of change were to remain popular and eventually were reconciled with an evolutionary perspective by arguing, as the English historian Edward Gibbon did in his *The History of the Decline and Fall of the Roman Empire* (1776–88), that each civilization flourishes and then dies but the torch of civilization is passed on to a still more successful later one. This view had already been expounded in the fourteenth century by the Arab historian Abu Zayd Abd ar-Rahman Ibn Khaldun (Rosenthal, 1967) and was pre-eminently championed in the twentieth century by the historian Arnold Toynbee in his 12-volume *A Study of History* (1934–61).

The revival of a knowledge of classical learning during the Renaissance thus familiarized scholars with a number of alternative views of human history, including cyclical and evolutionary ones. The moral and punitive power of religious authority, and perhaps the natural conservatism of scholars, were such that most of these ideas, as far as they were accepted at all, were fitted into the prevailing model of degeneration.

These combinations provided European explorers with various frameworks for interpreting the exotic peoples they encountered in the course of their journeys. Degenerationist theory, coupled with classical accounts, suggested that the most remote parts of the world might be inhabited by monsters, since there would be found those humans who had been cut off longest from the sources of moral guidance in the Middle East. While explorers failed to discover the expected unipeds or people who had their mouth and eyes embedded in their chests, encounters with cannibals in the West Indies and Brazil were thought to confirm the doctrine of degeneration. On the other hand, classical descriptions of a primitive golden age suggested that remote corners of the world might contain remnant 'terrestrial paradises' inhabited by kindly, innocently naked, and defenceless peoples. These were the sort of people Columbus believed he had encountered when he first reached the Caribbean. Early in the sixteenth century, Pietro Martire d'Anghiera specifically compared the aboriginal peoples of the West Indies with classical traditions of the golden age (Hodgen, 1964: 371).

Mastering Destiny

A major reorientation of thought began to occur in north-western Europe during the seventeenth century as that region emerged as the hub of a world economy. Economic development was actively being encouraged by kings and municipal corporations, and change was occurring rapidly enough for significant transformations to be evident in a single lifetime. This produced increasing confidence in the ability of human beings to manipulate their environment technologically and bring about social and cultural changes. For the first time European intellectuals began to argue that it was possible for humans to utilize their skills to create a better life for themselves. Their emphasis on human self-reliance contrasted with the traditional Christian belief that any human success depended on God's grace. Likewise the belief that the future could be better than any past contrasted with the degenerationist view that an ideal way of life had been divinely revealed to human beings at the beginning of time and that the most later societies could do was to try to live up, as far as possible, to that ideal. Apart from the speculations of a few classical philosophers, this was the first time in human history that human beings had asserted a belief in their own capacity to build a better world.

The earliest systematic exponent of this belief was the English philosopher and statesman Francis Bacon (1561–1626). He protested against the idea that the culture of classical antiquity was superior to that of modern times. He also urged his contemporaries to reject the authority that medieval scholars had ascribed to the learned treatises of antiquity as being the ultimate sources of scientific knowledge and to study the world about them rather than what had been written about it. Bacon's emphasis on an inductive methodology that encouraged observation, classification, and experimentation greatly impressed his countrymen and encouraged a new spirit of scientific enquiry that was reflected in the importance that the Royal Society of London, founded by Charles II in 1660, placed on these methods. Bacon's ideas were reinforced by René Descartes's (1596–1650) arguments that the laws governing nature were universal and eternal and that God existed apart from his Creation, which he had made to function without further intervention. A greater belief in natural order had already been encouraged in north-western Europe by Protestantism's tendency to dismiss miracles and the intervention of saints and other supernatural beings in human affairs.

More optimism about human creativity was also evident in France in the late seventeenth century in the so-called quarrel between the ancients and the moderns. The 'moderns' argued that human talents were not declining and hence Europeans could hope to produce works that equalled or surpassed those of the ancient Greeks and Romans (Laming-Emperaire, 1964: 64–6). While Walter Raleigh and many other writers in the early seventeenth century had still believed that the world was hastening towards its end, by the second half of that century many Western Europeans felt optimistic about the future of their societies (Toulmin and Goodfield, 1966: 108–10). They began to see themselves as shaping their own destiny rather than as being merely actors in a supernatural drama.

While classical philosophers had speculated about progress as one of many possibilities, the widespread optimism of the seventeenth century reflected what people saw happening around them. The reasons for optimism included the scientific revolutions of the sixteenth and seventeenth centuries, as manifested pre-eminently in the discoveries of Galileo Galilei (1564–1642) and Isaac Newton (1642–1727). Their findings had completely transformed the scientific understanding of the universe and how it operated. Perhaps even more important for ordinary people was the increasing application of scientific discoveries to the advancement of technology, which was an important factor promoting economic growth in Western Europe. Finally there was a widespread belief that the literary creations of English writers beginning in the reign of Elizabeth I and of French ones under Louis XIV were equal or superior to anything created in antiquity. Especially among the middle classes, who were benefiting from economic development, these changes encouraged faith in progress and that humans could be the masters of their own destiny.

A growing belief in progress also inclined Western Europeans to regard the ways of life of the technologically less-advanced peoples they were encountering in various parts of the world as survivals of a more primitive human condition and to interpret the chipped and polished stone artefacts found throughout Europe as evidence that people had once lived there who lacked knowledge of metallurgy. That these opinions were expressed as original observations from the sixteenth to the eighteenth centuries by numerous people who appear to have been unfamiliar with each other's work suggests that a growing belief in progress was slowly encouraging an implicitly evolutionary outlook. Most antiquarians, beginning with Michel Mercati in the late sixteenth century, drew parallels between European stone artefacts and the stone tools currently being manufactured by the aboriginal inhabitants of the New World and the Pacific islands. Indeed, it was these parallels that initially suggested the artificial, as opposed to the natural or supernatural, origin of the stone tools being found in Europe. In 1719 Dom Bernard de Montfaucon noted that French megalithic tombs contained polished stone axes but not metal ones, and on this basis attributed these tombs to a nation that lacked knowledge of how to work iron (Laming-Emperaire, 1964: 94). The sixteenth-century English writer John Twyne used ancient Greek accounts to argue that in the first millennium BC northern Europeans had lived in a primitive fashion that resembled how North American Indians were living in his own day. The Elizabethan artist John White, who had visited the Roanoke colony, did not hesitate to use his remarkable paintings of the Indians of North Carolina as the basis for a later series of speculative illustrations of the ancient Britons (Kendrick, 1950).

In a similar fashion, the seventeenth-century contract philosophers, Thomas Hobbes and John Locke, treated aboriginal societies as representing a primitive stage in the process by which the earliest human beings had banded together and begun to create a civilized style of life. For Hobbes ([1651] 1955: 82), primitive society was a stage when life remained 'solitary, poor, nasty, brutish, and short'. According to Locke ([1690] 1952: 29), early European societies resembled those which in his own time still prevailed among the American Indians: 'in the beginning all the world was America'. In Locke's opinion, these societies lacked the political authority necessary to secure the personal and property rights on which civilized society depended. As a result 'a king of a large and fruitful territory there feeds, lodges, and is clad worse than a day-laborer in England' ([1690] 1952: 25).

On the other hand, the sixteenth-century French essayist, Michel de Montaigne (1965: 150–9), echoing Tacitus, described the Tupinamba Indians of Brazil as having a rude style of life but one that was free of many of the evils and corruptions of the European upper classes. A similar view survived the French lawyer Marc Lescarbot's ([1617–18] 1907–14) prolonged and close encounters with the Micmac Indians of Nova Scotia early in the seventeenth century, while his English contemporary Thomas Hariot ([1588] 1951: 32) praised the ingenuity and intelligence of the Indians of Virginia, notwithstanding their primitive technology and lifestyle. By the 1640s, the French Jesuit missionaries living among the Huron Indians of Canada were to proclaim that these Indians' generosity and ability to cope with adversity far exceeded that of most Europeans, while denouncing their failure to beat their children as an omission that was conducive to gross depravity. One of them concluded that the North American Indians were more intelligent than French peasants (Thwaites, 1896–1901, 26:

125). Not everyone who viewed aboriginal societies as examples of what life had been like in Europe in earlier times was prepared to admit that a technologically less advanced way of life was a morally inferior one. This opinion laid the basis for the development of the so-called myth of the 'noble savage' that by the eighteenth century was to provide a romantic counterpoise to an evolutionary view of human history. The concept of the noble savage was an invention neither of the eighteenth century nor of the salons of Paris (Fairchild, 1928; Sioui, 1992).

The claim by so many different writers and for so many different reasons that the way contemporary primitive people lived resembled how at least some Europeans must have lived in the remote past suggests that by the late seventeenth century the rudiments of an evolutionary view of human history were already firmly established as a facet of Western European intellectual life. Yet advocates of progress were reluctant to claim that technological progress was a general theme of human history or to formally abandon a Christian belief in degeneration. Mercati and many other antiquarians argued that the early knowledge of metallurgy recorded in the Bible must have been lost by groups who migrated into areas where iron ore was not found. Others maintained that stone tools were used at the same time as metal ones by classes, communities, or nations that were too poor to own metal. Without adequate chronological controls, which archaeologists were unable to provide before the early nineteenth century, there was no factual evidence that would make an evolutionary view of human history seem more plausible than a degenerationist one. While many antiquarians were attracted to evolutionary interpretations of the existing evidence, they were reluctant to abandon completely a traditional degenerationist perspective on human history and to view all development as being from primitive to civilized. Starting with Acosta, an evolutionary outlook began to expand within the general context of degenerationism. Antiquarians accepted the biblical statements that the earliest humans had possessed all the key arts of civilization, but believed that some groups had lost them as they expanded into more remote regions of the world. A growing faith in the ability of human beings to reacquire or reinvent lost skills did, however, reflect increasing optimism about the creative capacities of human beings. Neither the Renaissance discovery that the past had been radically different from the present nor the clearer realization that development was occurring in contemporary Western Europe and therefore might have occurred elsewhere in the past led to the conclusion that progress was a general theme of human history. In part this caution may reflect the fact that many intellectuals were not prepared to outrage religious authority and public opinion by cultivating a view of history from which the need for divine intervention had been eliminated. Yet in Protestant countries extremely heterodox studies of human history were being published (Rossi, 1984). More likely it indicates that the question of whether there was an overall direction to human history, and what that direction might be, was not yet regarded as particularly important. In the seventeenth century successive periods tended to be viewed as a series of variations on themes that were grounded in a fixed human nature, rather than as constituting a developmental sequence worthy of study in its own right (Toulmin and Goodfield, 1966: 115–23).

3
Enlightenment Evolution

An evolutionary perspective on human history that was sufficiently comprehensive to challenge and replace the medieval view was formulated by French Enlightenment philosophers in the middle and late eighteenth century. Theirs was not a philosophy of contemplation but a call to action by people seeking to create a new social order. For the first time in modern history it was assumed that change was normal and resulted in general human betterment. Many of these philosophers were deists, who accepted Descartes's idea of a self-regulating universe and tried to understand human behaviour in naturalistic terms. They strove to effect reform by documenting in what ways existing conditions were antiquated, dysfunctional, and unjust, and by advocating changes that they believed were progressive and would benefit all humanity. By seeking to demonstrate that the general course of human history was marked by progress, they made their own demands for change seem to accord with the nature of things, and the long-term success of these demands appear inevitable.

Origins

The first Enlightenment philosopher was Charles-Louis, Baron de Montesquieu, whose *L'Esprit des lois* (1748) analysed relations between political structure, social organization, the economy, and other aspects of social life and offered a classification of different forms of government. The first comprehensive formulation of the concept of sociocultural progress was presented in 1750 by Anne-Robert-Jacques Turgot. Thereafter Enlightenment thought was propagated in France by François-Marie Arouet de Voltaire and the Encyclopedists. Marie-Jean de Caritat, Marquis de Condorcet, produced one of the culminating works of the movement, *Esquisse d'un tableau historique des progrès de l'esprit humain* (1795), which outlined the progress of the human race from barbarism to Enlightenment. Enlightenment thought also flourished in Scotland in the eighteenth-century school of 'primitivist' thinkers, which included Adam Smith, William Robertson, John Millar, Adam Ferguson, Henry Home (Lord Kames), and James Burnett, who as Lord Monboddo remains notorious for his tireless advocacy of the idea that human beings and orang-utans belong to a single species. Enlightenment philosophers sought to produce a theory that would explain social change and also to create a methodology that would enable them to study the general course of human history from the Creation to their own time.

However Enlightenment philosophy portrayed itself, it was a movement by which significant portions of the French and Scottish middle class sought to bring about reforms that would better serve their own collective interests. In England and the Netherlands, where political power was already in the hands of the commercial middle class, intellectual effort was directed mainly towards assessing the practical political and economic significance of this change and determining how political power might more effectively serve the interests of the middle class. The continuing weakness of the French entrepreneurial middle class under the Bourbon autocracy encouraged French intellectuals to engage in more broadly-based speculations about the nature of progress and to formulate arguments that identified their own class interests with progress in general (Hampson, 1982; Im Hof, 1994). The great impact that these ideas had on thinkers in Edinburgh reflected not only the close cultural ties between Scotland and France and the greater political power that had been acquired by the Scottish middle class following the Act of Union with England in

1707, but also a specific concern with how the Highland regions of the country could be made more productive. These backward zones, with their Gaelic-speaking and partly Roman Catholic population, had long posed a threat to Scottish unity. The challenge of developing the Highlands stimulated a special interest in the comparative study of societies and how change in general takes place (Crawford, 1992: 16).

Tenets

For Enlightenment philosophers it was not enough to know that societies changed; they wished to understand the direction and causes of such changes so that the middle class and, through the middle class, society as a whole might benefit from them. Enlightenment philosophy was constructed around a number of major tenets that were to become the basis of popular evolutionary thinking among the reform-minded middle classes throughout Europe and in America.

One of the most important of these tenets was the doctrine of psychic unity. Enlightenment philosophers believed that all human groups shared the same general kind and level of intelligence and the same basic nature, although it was recognized that individuals within groups differed from one another to some extent. Because of this, there was no biological barrier to the degree to which any racial or national group might benefit from new knowledge or contribute to its advancement. All human groups were equally perfectable and able to participate in the work of enlightenment.

Racism existed in the eighteenth century. In its most extreme form it was associated with the doctrine of polygenesis, which maintained that each human race had been created separately and constituted an immutable species. Enlightenment philosophy shared with orthodox Christianity the monogenist belief that all human groups were descended from a common ancestral couple who had been divinely created only a few thousand years ago, and that because of this these groups were similar in nature. Differences in physical appearance were explained as adaptations to particular environments that were thought to occur quickly and have little behavioural significance. Adverse climatic and other environmental factors might lower the intelligence of particular groups or predispose their members to greater violence and other undesirable forms of behaviour. But it was also maintained that such effects, even if they became biologically inherent and were passed on physically from one generation to another, could be reversed in the course of a few generations by means of education or a better diet.

Unlike Christian theologians, Enlightenment philosophers believed that human beings were basically good. This optimistic estimate of human behaviour implied that people's actions tended to be guided by reason and that they were prepared to seek the well-being of others alongside their own interests, so long as their true nature had not been corrupted by socially induced greed and pride. Being equipped with natural decency and good sense, human beings no longer believed that they had to depend on God or the Church to hold their evil tendencies in check. Enlightenment philosophy rejected medieval concerns with sinfulness and individual dependence on divine grace as the only means of achieving salvation. It therefore tended to encourage a secular humanism. Unlike the humanism of classical times, which contended that it was difficult, or even impossible, for individuals to defy the all-powerful fate that governed human lives, Enlightenment philosophers saw nothing preventing human beings from creating a better, or even a perfect, society (Bryson, 1945; Meek, 1976).

While Enlightenment philosophers believed that European civilization, and in particular their own ideas, represented the cutting edge of human development, they also thought that all humans could participate in this development. Napoleon, when he conquered Egypt in 1798, argued that life for the Egyptians could be transformed for the better by introducing Italian colonists, who would teach them more productive forms of agriculture, and by introducing French property laws. Asked whether he did not fear that some day Egypt might become strong enough to conquer France, he replied that, if that were ever to happen, it would be in the interests of the French to be conquered (Herold, 1962: 16). Even if this reply embodied much cynical political rhetoric, it accorded with a view of culture that, while markedly ethnocentric, was not racist. There was nothing in Enlightenment philosophy which suggested that advanced civilization was to remain exclusively European.

Enlightenment philosophers believed that progress was the dominant feature of human history and occurred continuously rather than episodically. They also saw progress bringing about not only technological but also social, intellectual, and moral improvement. The principal motivation for progress was the constant desire of individual humans to improve their condition, especially by gaining greater control over nature (Slotkin, 1965: 441). They did this through the application of universal human powers to observe the world around them and using their capacity to reason to devise new and more effective ways of doing things. In this fashion, human beings gradually acquired greater ability to control their environment, which in turn generated the wealth and leisure needed to support the creation of more complex societies and the development of a more profound and objective understanding of human affairs and of the natural world. The exercise of reason had, of course, long been recognized in the Christian tradition as a crucial feature distinguishing human beings from animals. It now became the key to individual and collective betterment. Many regarded progress as an invariable law of nature and therefore inherent in the human condition; others thought of it as something to be hoped and striven for.

Some Enlightenment thinkers went beyond this position to view social existence and social change as an autonomous process that imposed limits on human behaviour. The physiocrats, a school of economists advocating major tax reform that included François Quesnay, Victor Riqueti, Marquis de Mirabeau, and Turgot, argued that even absolute governments were helpless against the dynamics of social progress unless they understood and learned to obey the anonymous social forces that shaped the economy, population, and manners as an integrated whole. On the other hand, by understanding and learning to use these forces, governments might avoid decline. The physiocrats argued that, if the French monarchy was to survive, an enlightened administration had to be created that could govern in accordance with the natural laws of social process (Elias, 1978: 42–6). Their aim was to promote reform and advance middle-class interests by subjecting absolutism to the control of natural law. They further specified that beneficial social change could only come about as the result of a better understanding of how society works.

In these early theorizings about the nature of sociocultural change we can discover alternative positions that have persisted to the present. On the one hand, increasing knowledge is hailed as liberating human beings from the direct control of nature and enriching their lives. On the other, success depends on recognizing and taking account of the influence of social laws that exist independently of human volition and constrain human action. While both views identify knowledge as the basis of effective and predictable action, a vision of human liberation is set against a vision of continuing constraint. While not mutually exclusive, these two sorts of explanation appeal to different sorts of social scientists.

Enlightenment philosophers also believed that progress characterizes not only technological development but all aspects of human behaviour, including social organization, politics, morality, and religious beliefs. Changes in all these spheres were viewed as occurring concomitantly and as following, in a general fashion, a single line of development. As a result of similar ways of thinking, human beings at the same level of development devise similar solutions to their problems; hence, even in the absence of any direct or indirect contact, their ways of life tend to develop along parallel lines. Cultural change was frequently conceptualized as following a universal series of stages; with Western Europeans having evolved through more stages than anyone else, while technologically less advanced societies had passed through only the elementary ones. It was Mirabeau, in the 1760s, who first used the term 'civilization' to designate the highest of these stages (Elias, 1978: 38–40). This process is commonly called unilinear evolution. Yet, since it assumes that societies in many parts of the world and often at different times pass through the same sequence of stages, the term parallel evolution seems preferable. Neither in the eighteenth century nor later did parallel evolutionists rule out diffusion and migration as additional processes by which innovations might be propagated. They had sufficient confidence in psychic unity and human inventiveness, however, to believe that in the absence of contact different groups of people would independently devise similar solutions for the same problems.

Enlightenment philosophers were not biological evolutionists and their ideas about cultural change did not negate the traditional notion of human nature as something that was fixed and immutable. Progress was claimed to perfect human nature not by changing it but by progressively curtailing dangerous passions and eliminating ignorance and superstition. By acquiring more accurate knowledge about the nature of things, people liberated themselves not only from domination by nature but from antiquated and debilitating customs, institutions, and taboos. This freed humans to forge a new future on the basis of reason, natural justice, and scientific knowledge.

Although often deists, who believed that God existed apart from his Creation and that human beings were responsible for working out their own problems, most Enlightenment philosophers viewed cultural progress teleologically. They believed that God had endowed both humanity and the universe with innate qualities that promoted progress and human well-being. If God was no longer thought to intervene directly at crucial points in individual lives and to change the course of history, cultural progress still represented the realization of the plans of a benevolent deity. A faith that such benevolent laws guided human development would long outlive a concern with God among social scientists (Gellner, 1988: 144).

Conjectural History

Eighteenth-century French scholars were familiar with the ideas of Lucretius and knew that archaeological evidence indicated that stone tools had once been used throughout Europe. They were also familiar with classical and biblical texts which suggested that bronze tools and weapons had been used prior to iron ones. In 1734 Nicolas Mahudel read a paper to the Académie des Inscriptions in Paris in which he set out the idea of three successive ages of stone, bronze, and iron as a plausible account of cultural development. In 1758 Antoine-Ives Goguet supported the three-age theory in a book that was translated into English in 1761 as *The Origin of Laws, Arts, and Sciences, and their Progress among the Most Ancient Nations*. He believed that modern 'savages set before us a striking picture of the ignorance of the ancient world, and the practices of primitive times' (Heizer, 1962: 14). Enlightenment philosophers did not, however, turn to archaeology as a source of information about the past. As yet, in the absence of written records, antiquaries could do little to arrange prehistoric finds in chronological order. Moreover, stone tools and broken pots seemed unlikely to reveal much that would interest philosophers, who were anxious to learn more about the thoughts and behaviour of human beings in remote times.

In place of archaeological data, Enlightenment philosophers used the method that Dugald Stewart labelled 'theoretic' or 'conjectural' history to trace the development of human institutions (Slotkin, 1965: 460). This involved the comparative study of living peoples whose cultures were judged to be at different levels of complexity and arranging these cultures to form a logical, usually unilineal, sequence from simple to complex. Most of the data used in these studies were derived from accounts by missionaries and explorers of peoples living in many different parts of the world, although Greek and Roman accounts of less developed peoples some 2000 years earlier were also utilized. Despite disagreements about significant details, such as whether agricultural or pastoral economies developed first, it was believed that these sequences could be treated as if they were historical ones and used to examine the development of all kinds of social institutions from their beginnings to their currently most evolved state. It was usually argued that pastoral societies had preceded agricultural ones, since pastoral economies were less intensive and supported lower population densities.

In general, the French were most interested in delineating moral and intellectual progress, while the Scots tended to focus on economic development (Burrow, 1966: 10). In these schemes, which generally ran from hunter-gatherer societies through small-scale agricultural ones to pre-industrial civilizations, such as those of China and India, modern non-Western cultures were used to illustrate what European societies had been like at various periods of antiquity. The total range of these cultures, from simplest to most complex, represented the successive stages through which European civilization had evolved in the course of its history.

This use of ethnographic data made sense only within the framework of parallel evolution. It also reduced modern, non-Western societies to the status of 'living fossils', whose main value was their illustration of how Europeans had lived in the past. This approach also encouraged a growing emphasis on documenting the primitive and often the most brutal characteristics of non-Western peoples in order to marshal them into evolutionary schemes. That discouraged older tendencies to see supposedly less advanced cultures as a mixture of primitive and more advanced traits.

The demonstration by the Scottish historian William Robertson and others that a similar range of societies could be found in both the Old World and the Americas was interpreted as evidence that human societies had developed independently along similar lines in both parts of the world, and this in turn was believed to validate the assumption of parallel evolution. In the nineteenth century, additional support for this sequence was seen in the shading of one stage into another and the survival into later stages of elements of behaviour that appeared to have arisen at an earlier stage (Burrow, 1966: 13–14).

In spite of these arguments, it was still widely believed in the eighteenth century that the aboriginal peoples of the Americas and most Old World groups had regressed to the status of hunter-gatherers before they began their climb towards civilization. As late as 1862, Daniel Wilson (1862, 1: 144, 183) was to describe the Stone Age not simply as an early stage in human development, but as a base level to which individual human societies occasionally declined and from which they then had to reascend. What Enlightenment philosophers sought to document was the natural tendency for human societies to progress and that the overall course of human history was in an upwards direction. The physiocrats hoped that through a better understanding of economic laws they might arrest the tendency of particular regimes, particularly that of France, to decline as a result of corruption and misadministration. There was much less interest in engaging in disputes with religious authorities concerning the original state of humanity.

One of the main challenges evolutionists faced was to account for why all peoples had not progressed at the same rate. The usual answer given at this period was an environmental one. Since classical times, it had been believed that the intelligence, personality, and creativity of different human groups were influenced by the environmental settings in which they lived. These influences were exerted through climate, diet, and the varied potentialities for development of different regions. Environmental changes were believed to bring about biological changes in human beings very quickly as people moved from one region to another. Noting not only the relative cultural backwardness of the native peoples of the Americas, but also the relative smallness and weakness of American plants and animals compared with their European and Asian counterparts, a number of eighteenth-century naturalists and historians, including William Robertson, argued that the climate of the New World was conducive to the degeneration of plant and animal life. They also suggested that over the course of a few generations this climate might adversely affect European settlers (Haven, 1856: 94). While extreme environmental limitations clearly explained why the Inuit had remained hunter-gatherers, environmental explanations of different rates of cultural evolution for most regions of the world required invoking factors that today seem improbable and were probably not fully convincing when they were first proposed. These included the belief that damper and hotter climates were less favourable for achieving the higher stages of cultural evolution than were drier and more temperate ones.

Archaeological data began to supply independent support for the theory of sociocultural evolution only in the first half of the nineteenth century, when Christian Thomsen's seriation technique and Jens Worsaae's stratigraphic excavations demonstrated that there had been a succession of stone, bronze, and iron ages in northern Europe. Yet the Scandinavian archaeologists interpreted their finds not as a local development but as the consequence of new peoples and technologies reaching Scandinavia from the south (Gräslund, 1987; Klindt-Jensen, 1975). As late as 1888, the Canadian geologist John William Dawson was able to argue that the prehistoric developmental sequence found in Europe might represent only local trends or the accidental interdigitation of contemporary groups that had different cultures. He concluded that there was as yet no clear evidence that cultures at various levels of complexity had not co-existed throughout all human history. It was easier at that time for sociocultural evolutionists to ignore Dawson's objections than to refute them. Not enough was known about prehistoric sequences outside northern and western Europe to demonstrate archaeologically that sociocultural evolution had been a general trend throughout human history.

Impact

Enlightenment philosophers evolved a general theory of sociocultural evolution, which they supported with comparative ethnographic evidence, long before any conclusive archaeological data were available to sustain such a theory. Although the possibility of biological evolution, or transmutation, was considered by Montesquieu in 1721, the French mathematician Pierre-Louis Moreau de Maupertuis and the Encyclopedist Denis Diderot in the 1750s, and the naturalist Georges Buffon in the 1770s, it only began to be examined seriously by Erasmus Darwin in 1794 and by Jean-Baptiste de Monet, Chevalier de Lamarck in 1809. It is therefore clear that the concept of sociocultural evolution developed prior to and independently of that of biological evolution. Lamarck's theory of the biological inheritance of acquired characteristics was a transposition of ideas about cultural change into the biological realm, although it took a long time for the errors involved in doing this to be recognized (Steadman, 1979: 129).

Enlightenment philosophy served the interests of the commercial middle class but also reflected a growing awareness of the irreversible changes that were transforming every aspect of Western European society and culture. By suggesting that the future would be better than the past and that the past did not provide a model for the present, French sociocultural evolutionists forged a powerful basis for legitimizing their critiques of existing institutions and demanding that they be reformed. In the latter half of the eighteenth century, existing institutions had to justify themselves in relation to the twin criteria of reason and serving the best interests of society and the individual. Any institution which stood in the way of entrepreneurship was called upon to justify itself before the bar of history. This was a challenge to which the beneficiaries of the *ancien régime* found themselves unable to respond.

4
Romantic Reaction

Although it developed primarily in Paris and Edinburgh, Enlightenment philosophy stimulated new ways of thinking throughout continental Europe, Britain, and America, especially among the middle class and the more liberal members of the aristocracy. Its evolutionary perspective enabled exponents everywhere to call into question traditional institutions which they saw standing in the way of self-improvement and social reform, while its rationalism and emphasis on human goodness encouraged people to believe that it would be relatively easy to create a new social order that would stimulate economic development and promote social justice. Although initially aiming to ameliorate rather than destroy existing governments, Enlightenment thought became an important catalyst in precipitating the American and French revolutions and in shaping the republican regimes to which they gave rise.

Critique of the Enlightenment

It is not surprising that in the United States, where Enlightenment views played a major role in creating a highly successful new political order, these ideas remained popular into the early decades of the nineteenth century. During this period Enlightenment thought provided the intellectual basis for the new republic. In Europe, however, Enlightenment thought quickly became abhorrent to established groups who found themselves threatened by demands for social and political reform. In particular, the French Revolution stimulated a conservative backlash that was directed against not only the revolutionaries but also the ideas that had inspired them. European society became divided into two camps: those who hoped to benefit from the spread of French republicanism and those who saw their interests threatened by it. French armies carried the revolution into Italy, Germany, and the Netherlands. In Germany and Italy, at least initially, there was much support for the revolution and for Enlightenment thinking among the hitherto politically suppressed middle class. In England, where the middle class already enjoyed political power, the French Revolution was feared as potentially encouraging an uprising among the working class. Middle-class intellectuals, such as Erasmus Darwin, who had hitherto been receptive to Enlightenment philosophy, sought to escape establishment reprisals by espousing more conventional ideas (Desmond and Moore, 1992: 10–11). To uphold Enlightenment beliefs was equated with encouraging anarchy, bloodshed, and republicanism. The same stigma was not attached to Enlightenment ideas in Scotland, where these beliefs had been espoused by the establishment and were already dissociated from political radicalism. While Scottish thinkers remained fascinated by an Enlightenment interest in questions of origins, English social analysts continued to invoke Baconian inductivism as a reason for studying how society worked rather than speculating about how it might have evolved.

The reaction against Enlightenment philosophy was not, however, directed simply against its political radicalism. Deficiencies in Enlightenment philosophy had been noted long before it had assumed an active political role and most of the early criticisms came from within the movement. Jean-Jacques Rousseau saw the rationalism of the Enlightenment as failing to take account of human feelings, emotions, and sensations, which he rightly regarded as important aspects of human experience and major spurs to action. Rousseau stressed the importance of maintaining close contacts with the natural world and of developing human sensibilities. While he believed in the inherent goodness of human beings, he observed that people were corrupted all too easily by civilized society and that progress in learning and economic betterment often led to inequality, exploitation, and feelings of alienation rather than to the improvement of human beings. While Rousseau accepted that the various parts of sociocultural systems were interconnected, he rejected the Enlightenment belief that progress in one aspect of social life inevitably led to improvement in others. Like Tacitus, he believed that economic progress and growing material and political inequality resulted in social and moral degradation.

Others reacted against the universalism of the Enlightenment, which treated all human beings as equal but tended to ignore or slight the distinctive cultural traditions and the family, ethnic, and regional allegiances that shaped and gave meaning to the lives of most people. Both before and after the French Revolution, there was growing emphasis on local cultural traditions as a reaction against French cultural domination and the literary and artistic restrictions of a classicism that had become identified with both France and the Enlightenment.

The Romantic Reaction

An emphasis on sensibility and feelings, including those of the most bizarre and extreme sort, as well as on local cultural traditions gave rise to the romantic movement, which was initially centred in the German states and started as a middle-class protest against the French culture of the German ruling class (Dumont, 1994).

Romanticism began with the German literary revival of the eighteenth century, which took the form of a revolt against classicism and sought to glorify Germany's ancient and medieval past (Hampson, 1982: 239–50). This movement was provided with a philosophical basis in the 1770s by the philosopher Johann Herder, who defined history as the account of the development of individual peoples as exemplified by their languages, traditions, and institutions. Herder stressed the importance of individual cultures, which he viewed as the embodiment of the creative energies of peoples who shared the same language and the same patterns of thought in their everyday life. True culture was the repository of the shared beliefs of ordinary people: townsmen, farmers, tradesmen, and intellectuals. It was not the cosmopolitan affectations of the aristocracy (Barnard, 1965; Beiser, 1992). Like most Germans, who lived in a patchwork of principalities and small kingdoms, Herder saw no necessary connection between nations and states.

Contrary to the Enlightenment philosophers, Herder argued that one culture was not inherently superior or inferior to another. Each was an expression of humanity's potential to create many different patterns for living. Societies were viewed as organic wholes, in relation to which the thoughts and actions of individuals acquired their meaning. Simpler cultures had to be understood on their own terms and could not be treated simply as illustrations of stages through which more advanced ones had evolved in the past. In general, German scholars tended to view cultures not as evolving slowly over time, but as already perfectly formed at an early stage. Such progress as did occur represented the unfolding of the spiritual potential that was latent in each culture (Bowler, 1989: 70). True insight into a culture was found by seeking to reconstruct in one's own imagination the spirit that had animated it.

Herder and the German intellectuals who followed him tended to stress the extraordinary cultural diversity of humanity rather than those features that all human beings shared as members of a single species. In keeping with this view, the German anthropological tradition was to emphasize the idiosyncratic and unpredictable nature of change, while the evolutionary approach that was derived from the Enlightenment would stress cross-cultural similarities and parallelisms in cultural development. In the nineteenth century, German culture-historical anthropologists would study the development of many individual cultures, while evolutionists tended to study the evolution of culture as a unitary, pan-human phenomenon. Lacking a unified state prior to 1870, middle-class Germans sought to find their common identity in a distinctive culture. Feeling politically and economically inferior to the French and British, they took refuge in the idea of the inner worth and incomparability of each culture as an expression of the human spirit.

Many scholars, including Marxist ones, have viewed the philosopher Georg Hegel's idealist view of human history, as a process by which humanity achieves spiritual and moral progress and attains greater self-knowledge, as being an evolutionary one. Yet Hegel's assumption that humans will eventually recognize themselves as spiritual and as one with God conceptualizes humanity as moving in accordance with a divine plan towards the regaining of the Absolute, which is the all-embracing totality of reality, through the negation of the former negation that had constituted the finite, differentiated universe. While Hegel may have conceived of human history as a developmental process, this mystical and romantic philosopher was thinking in a fashion that was essentialist, archetypal, and cyclical rather than evolutionary.

Both the emotional and nationalistic aspects of romanticism influenced literature and music throughout Europe. Writers tended to draw inspiration from old legends and ballads and to search for unrecorded survivals of medieval epic poetry, or even invent it when genuine material was not forthcoming. Music drew increasingly on local folk tunes. Romanticism rejected efforts to limit and control human experience and reached out to encompass the exotic and the bizarre. Romantic literature exploited the full range of subjective states. The Gothic novel mixed tales of horror and evil with poetic descriptions of natural beauty. Romanticism recognized that there were irrational and even violent aspects to human behaviour and expanded the compass of art to include a much broader range of behaviour than rationalism and classicism had accepted as appropriate. It also recognized evil as well as benevolent tendencies in human beings, and self-destructive as well as life-affirming attitudes. It tended to idealize primitive or natural societies and treated the spirit of European nations of the medieval and still earlier migration periods, as preserved in their oral traditions, as an inspiration for modern arts and letters. The violent and uninhibited behaviour of people in these early periods was regarded as a healthy contrast to the inhibitions and self-control of modern Europeans.

Romanticism and Nationalism

Romanticism acquired additional importance in the context of the conservative restoration and the concerted effort to crush Enlightenment ideas that followed the defeat of Napoleon. The claim of the Enlightenment that progress benefited everyone was widely denounced as a shallow excuse for promoting French cultural domination. The nationalism that had been stimulated by the Napoleonic Wars continued to grow, strengthening the nation states of Western Europe and threatening the unity of the multinational dynastic regimes of Eastern Europe, in particular those presided over by the Austrian, Russian, and Turkish monarchs. Romanticism helped to promote national pride by recording traditional culture, popularizing and glorifying it, and whetting an interest in tracing its origins and history. Sir Walter Scott's colourful portrayals of Scotland's past, beginning with his novel *Waverley* (1814), inspired imitators in all European countries, and this in turn was an incitement to romantic nationalism. At the same time there was a growing belief that alleged differences in national temperament and behaviour were inherent in human biology. This suggested not only that ethnic groups were biologically as well as culturally different, but also that differences in their character could not be changed easily or quickly. This claim was a direct challenge to the Enlightenment belief in psychic unity and to parallel evolution.

Nationalism tended to focus and intensify diffuse racial sentiments that prior to the nineteenth century had been constructed along class lines as often as along geographical ones. Prior to the revolution, French aristocrats commonly boasted that they were descended from German-speaking Franks who had invaded France and conquered its indigenous Romano-Gallic inhabitants. The descendants of the original population of France thus constituted the middle and lower classes. In Britain the aristocracy likewise traced its descent from the Normans, who had conquered that country's earlier Anglo-Saxon and Celtic inhabitants in AD 1066. In Poland the aristocracy claimed to be descended from the original Poles who had subdued and imposed their language on their country's original non-Polish inhabitants. At the time of the French Revolution, the suppression of the aristocracy was often described as the expulsion of foreign conquerors and the restoration of sovereignty to France's original inhabitants (Dietler, 1994). In each of these cases, ethnic and biological differences were used to enhance and explain class differences.

In revolutionary France, however, under the influence of the Enlightenment, priority was accorded to nationality being based on citizenship rather than on ethnic origins. Hence German-speaking Alsatians, as a result of their oath of allegiance to the new republic, became as French as anyone else. Nevertheless, the French authorities sought to ensure national unity by using the educational system and cultural institutions to promote the French language and revolutionary culture at the expense of ethnic diversity in regions such as Alsace and Brittany. In general, national identity was to be equated with linguistic and cultural unity throughout Europe (Anderson, 1983; Gellner, 1983). Yet the Enlightenment conviction that everyone is an individual human being first and French second became and has remained a vital part of French culture (Dumont, 1994: 199–201).

In Germany and many other parts of Europe the idea was accepted that nationality was also grounded in ethnicity and hence involved a common origin and a shared physical make-up, often metaphorically symbolized by shared blood. Cultural romanticism encouraged Germans to see themselves as part of the human family through being German rather than as a result of being individual human beings (1994: 199). This strengthened the belief that much of what is different between one nationality and another was biologically determined and could not be fundamentally altered by re-education or as a result of political or economic change. It directly contradicted the Enlightenment belief that the most significant differences among groups were related to their respective levels of development and therefore would be erased by progress. The belief that each ethnic group was distinguished by immutable biological differences that could not be altered by changing circumstances provided conservatives with a reason to reject social change as difficult, ineffectual, and contrary to human nature. They were also comforted by the belief that no matter how much social change occurred, individual peoples would remain fundamentally the same. Thus a racialized romanticism evolved as the doctrine of choice of political conservatives.

Romanticism and Racism

Racial ideas developed slowly during the first half of the nineteenth century but became much more important in the 1850s. Their most influential exponent was Joseph-Arthur, Comte de Gobineau, who published his 4-volume *Essai sur l'inégalité des races humaines* between 1853 and 1855. A member of an aristocratic and royalist French family, Gobineau exalted the Frankish nobility as the true creators of Western civilization (Bowler, 1989: 109). He believed that the fate of civilizations was determined by their racial composition and that the more a successful civilization's racial character became diluted through intermarriage with other groups, the more likely it was to sink into stagnation and decay. Gobineau's writings were to influence European racists from the nineteenth-century composer Richard Wagner to Adolf Hitler. Yet he was not alone. In 1850 Robert Knox had proclaimed in his *The Races of Men* that race and heredity were everything. He, like Gobineau, regarded racial hybrids, which he maintained included the children of English and Irish parents, as unnatural. He also subscribed to older ideas that races were unlikely to succeed in alien territory and doubted that without constant immigration European settlers could survive in North America and Australia (Bowler, 1989: 109–10). In the United States racism in the form of polygenism had been used since the 1830s to account for the alleged inferiority of blacks and Indians (Stanton, 1960). Later in the century European-style ethnic racism was employed to argue that the growing numbers of immigrants from southern and eastern Europe posed a biological threat to the fitness of the white race in America and to America's democratic institutions. These views were popularized by works such as Madison Grant's *The Passing of the Great Race* (1916). In both Europe and America novelists, popular writers, and reputable scholars were invoking racial factors as well as environmental ones to explain variations in the degree to which different groups had evolved in the course of human history.

As a result of growing interest in biological explanations of human behaviour, race replaced language as the main criterion that was used to trace the history of ethnic groups. It was erroneously assumed that, while people might easily learn a new language, in early times ethnic groups tended to be endogamous and hence racial types had persisted despite migrations and conquests. Racial theorists also claimed that only invasions and conquest could introduce a superior culture to a region (Bowler, 1989: 122–4). As cultures came to be viewed as biologically inherent in different populations, change was devalued and old customs were treated as more authentic reflections of a people and their culture than were recent innovations or borrowings. The development of racism provided an explanation for cultural variation that had great appeal to romantics because it identified the culturally specific and the exotic as the most essential reflections of human nature and offered reassurance that these aspects of human behaviour and identity were relatively resistant to change.

Revival of Degenerationism

Opposition to the Enlightenment doctrines of progress also came from another direction. In 1724 the French Jesuit missionary Joseph-François Lafitau, who had worked among the Indians of Canada, published his *Moeurs des sauvages amériquains comparées aux moeurs des premiers temps*. Although this book is often erroneously described as an early contribution to evolutionary anthropology, it was in fact a defence of the doctrine of cultural degeneration. Lafitau directed his attack against the deists, whose beliefs he saw as a menace to Christianity. Lafitau demonstrated many ways in which the beliefs and customs of the American Indians resembled those of the ancient Greeks and Romans, but he did not interpret these as evidence of parallel development in the eastern and western hemispheres as Enlightenment thinkers were to do. Instead he argued that they were corrupt and distorted versions of the true religion and morality that God had revealed to Adam and his descendants in the Middle East. The Greeks and Romans had carried these beliefs west and the ancestors of the American Indians had transported them far to the east. Along the way, much of God's revelation had been lost and what survived did so in a fragmentary and corrupted manner and in forms that were different among the ancient Greeks and Romans from those found among the American Indians. Throughout the eighteenth century, conservatives could point to Lafitau's work, with its diffusionary–migrationary explanation of world history, as a respectable alternative to the idea that similar beliefs and customs were invented repeatedly as part of the process of cultural development in various parts of the world. Without detailed archaeological data, it was impossible to determine which of these two versions of world history was the more realistic.

Degenerationism became linked with early tendencies toward romanticism in the writings of the eighteenth-century English antiquarian William Stukeley. Stukeley was a perceptive observer of ancient monuments and one of the first English antiquarians who recognized the possibility of a lengthy pre-Roman occupation of Britain, during which distinctive types of monuments had been constructed at different times and different peoples might sequentially have inhabited England. Stukeley associated all the major Neolithic ritual centres in Britain, including Stonehenge and Avebury, the latter of which he investigated in particular detail, with the druids, a Celtic priesthood whose existence had been noted by the Romans. Stukeley believed that the druids had brought with them to England from the Middle East knowledge of the original, divinely revealed religion. He also maintained that this religion had survived in a relatively pure form in Britain, as a result of which druidical beliefs and practices were closely akin to those of Christianity. Like Lafitau, Stukeley interpreted parallels among religions as evidence that they sprang from a common source, rather than as proof that peoples at similar levels of development interpret the world in a similar fashion, as Enlightenment philosophers were to do. Like Lafitau, Stukeley was defending traditional Christian beliefs against the deist claim that reasonable people can apprehend God without the help of revelation (Piggott, 1985; Ucko et al., 1991: 74–98). Stukeley's interest in linking Britain in a romantic fashion with the development of the Judaeo-Christian religious tradition persisted into the late eighteenth century, most prominently in the work of the artist and mystic William Blake.

Rationalism and Romanticism in the Nineteenth Century

The British archaeologist Stuart Piggott (1985: 115–17, 154–5) has viewed the late eighteenth century as a period of marked intellectual decline in historical and antiquarian studies in England and attributed this to a general slackening of scientific rigour that was brought about by romanticism. Romantics were interested not in the ultimate origins of human culture or in universal history, but rather in the histories of specific peoples. This attitude discouraged the study of cultural evolution. Yet, contrary to Piggott, it did not discourage an interest in archaeology. Romantics were keenly interested in ruined abbeys, ancient graves, and other symbols of death and decay, including human skeletons grinning a 'ghastly smile' (Marsden, 1974: 18). They also saw the spirit of European nations preserved in their monuments as well as in their folk traditions. This was instrumental in encouraging the excavation of ancient sites, especially burials. From the 1750s on, hundreds of early historical and prehistoric graves were excavated in Britain. James Douglas's *Nenia Britannica, or Sepulchral History of Great Britain*, which was published in parts between 1786 and 1793, was based on information recorded in the course of burial excavations throughout the country. Never before had so much archaeological work been accomplished in Britain, nor had so much of it been recorded with such care. But in Britain the repudiation of Enlightenment philosophy in the late eighteenth century led to the abandonment of an interest in sociocultural evolution. Archaeology focused on the Roman, Saxon, and medieval periods more than on prehistoric times (Van Riper, 1993: 39–43).

In Denmark archaeology developed differently. There the struggle of the middle class against absolutism encouraged a continuing interest in Enlightenment ideals, including cultural evolution. Yet it was nationalism, spurred by the long-remembered loss of Skane to Sweden in 1658 and the British naval bombardments of Copenhagen in 1801 and 1807, that promoted a romantic interest in the Danish past and the amassing of a large prehistoric collection as part of a Museum of Northern Antiquities. If the evolutionary idea of successive ages of stone, bronze, and iron encouraged Christian Thomsen to devise his scheme for using stylistic criteria to seriate prehistoric archaeological finds into a chronological sequence, it was nationalism and a romantic interest in the past that had brought the collection he was studying into existence. Hence prehistoric archaeology came about as the result of a combination of romantic and Enlightenment interests.

In the reactionary years that followed the final defeat of Napoleon in 1815, romanticism became dominant and the ideals of the Enlightenment went into eclipse everywhere in Europe, surviving mainly in the form of radical protest movements among the lower middle and working classes (Desmond, 1989). The main exponents of an Enlightenment viewpoint in the early nineteenth century were the French social reformer Henri de Saint-Simon and Auguste Comte, the founder of positivism and sociology. Saint-Simon foresaw the importance of industrialization and believed that science and technology together could solve most social problems. He urged property-owners to defend their leading position in society against the threats of the propertyless by subsidizing the advance of knowledge. That would guide society toward improving the living conditions of the lowest classes as quickly as possible. In the tradition of the Enlightenment philosophers, Comte maintained that progress was inherent in human nature and that human intellectual development carried humanity from a theological stage, when human destiny was thought to be controlled by the actions of gods and spirits, through a transitional metaphysical stage, in which explanations were framed in terms of essences, final causes, and other abstractions, to a final positivist phase, when humans at last became aware of the limitations of their own knowledge. Comte argued that discovering regular connections among phenomena was the only basis for prediction and therefore for effective action. In his later work, he treated moral progress as the central goal of human effort and sought to delineate the sort of political organization that such progress required. Saint-Simon and Comte maintained an Enlightenment optimism about the essential goodness of human nature and the power of reason that was to persist into the twentieth century only in the Marxist tradition.

5
Racist Evolution

By the middle of the nineteenth century Britain had become the 'workshop of the world'. The industrial revolution, which had transformed an agrarian, handicraft economy into one characterized by the ever-increasing use of inanimate power sources, machine manufacture, and factory production, had begun in England during the eighteenth century. The first practical use of a steam engine, to pump water, had been made at the end of the seventeenth century and by the early nineteenth century steam engines were being used to power looms and other machines and to transport goods and people more efficiently than ever before. During the twentieth century internal-combustion engines and electricity have transformed communications, production, and everyday life at an ever-accelerating rate. If an Egyptian pharaoh had revived in seventeenth-century England, he might have been surprised by the extent to which wind- and water-powered machines were being used, but nothing about technology or society would have been totally beyond his comprehension. A pharaoh who revived in the late nineteenth century, and even more so in the twentieth, would have imagined himself in the realm of the supernatural. Technological change has not only occurred with increasing speed but transformed every aspect of social life and challenged traditional understandings (Adams, 1996).

As a result of these changes, a new industrial segment of the middle class assumed a position of leadership in relation to the established commercial and professional middle class. The rapid expansion of industry in the first half of the nineteenth century greatly reinforced the political power and self-confidence of the middle class as a whole, who came to view themselves as a major force in world history. An unprecedented pride in economic and social progress was apparent at the Great Exhibition held in London in 1851. It was in the context of these changes that during the middle of the nineteenth century England became the major centre of theorizing about biological and sociocultural evolution.

Evolution and Middle-class Power

Earlier in the nineteenth century, the concept of evolution had been a disreputable one in England. Lamarckian evolution, which was widely construed to imply that both Christian morality and the existing social order had spontaneously evolved from bestial proto-types rather than being part of an immutable divine dispensation, was associated with revolutionary politics and shunned by the respectable middle class. The latter saw their future and England's to be linked to the development of free enterprise and free trade within a social order protected by Christian dogma (Desmond, 1989). Fears concerning social stability were exacerbated in the 1840s by the Chartist movement and by the growing political unrest on the continent that culminated in the rebellions of 1848. Authors such as Benjamin Disraeli, in his novel *Sybil* (1845), and Frederick Engels, in *The Condition of the Working-Class in England* (1844), documented the growing class divisions within a society that was rapidly urbanizing and industrializing. Yet, by the 1850s, increased social stability and political compromises had calmed fears of mass insurrection and this once again disposed the middle class to view cultural change from an evolutionary perspective. Sociocultural evolution came to signify that the recent changes that had favoured the middle classes were neither a historical accident nor simply a form of self-aggrandizement, but the latest expression of an ongoing process that had shaped the whole of human history and promoted the welfare of all humanity. Sociocultural evolution became still more attractive when it was attributed to individual initiative, which occurred most effectively in the context of *laissez-faire* politics. Thus a sociocultural evolution that justified a non-interventionist form of government became closely aligned with liberal politics and the interests of the industrial middle class (Burrow, 1966: 265–9).

The first Victorian popularizer of a view of evolution that differed from that of the Enlightenment was the Scottish publisher Robert Chambers, who anonymously brought out *Vestiges of the Natural History of Creation* in 1844. This work went through numerous revisions and was widely read in England. Chambers believed that evolutionary processes were inherent in nature, although he did not propose a coherent mechanism to explain how they operated. Yet, by treating everything, from the development of life to the gradual perfecting of human society in deistic terms, as the natural unfolding of God's plan for the cosmos, Chambers rescued the concept of evolution from the domain of radical politics and began making it respectable among the established liberal middle class (Bowler, 1989: 91). At a time when the rapid increase and wide diffusion of scientific knowledge were weakening a belief in God's direct intervention in human affairs, this embedding of divine purpose into the structure of both human and cosmic history reassured a broad cross-section of the middle class that order and meaning were inherent in the universe.

The great exponent of evolutionary concepts in the 1850s was the philosopher Herbert Spencer. He argued that the cosmos, plant and animal life, and human society had evolved in that order from simple, homogeneous beginnings into increasingly differentiated, more complexly organized, and more intricately articulated entities. In 1857 he proposed that the pressure of feeding an expanding population spurred individual initiative and encouraged economic development. Societies that were more complex and better integrated were able to prosper at the expense of less complex ones, just as human individuals and groups who were better adapted to social life supplanted those who were less well adapted (Peel, 1971; Sanderson, 1990: 26–7).

Spencer believed that the key to success was individual freedom combined with the protection of property rights, which permitted people to enjoy the fruits of their personal initiative. Over time, the benefits of such unfettered individual enterprise made societies more prosperous and freer from coercion by natural forces or human tyrants. Spencer also adopted Lamarck's view that the characteristics that living plants and animals acquired in the course of their lifetimes could be passed on to their descendants. He was therefore advocating a similar view with respect to both biological and social evolution, although he recognized that acquired characteristics were transmitted differently in the sociocultural realm from how they were transmitted biologically. Like Comte, Spencer held industrial societies to be inevitable and desirable, and lauded thrift and hard work as leading humanity toward an earthly state of perfection. By praising individualism and free enterprise as the key factors bringing about progress, Spencer helped to make the concept of evolution acceptable to the middle class (Burrow, 1966: 220–5; Bowler, 1989: 91–2). In so doing, he encouraged all but the religiously most conservative members of the middle class to consider arguments favouring biological evolution and the possibility of a non-creationist origin for the human species. These concepts had formerly been anathema because they were seen as undermining religion and therefore the foundations of public order.

Some sociocultural evolutionists, especially those associated with political economy and philosophy, continued to write in the tradition of the Enlightenment. Many viewed evolution in a mystical, semi-Hegelian fashion as the spiritual development of humanity on its way to a universal realization of its true nature and celebrated the capacity of individuals to work together to create a better-educated and morally superior world (Melleuish, 1995: 150–1). Often this process involved liberating the individual from the tyranny of social collectivities, and attempts to create rational social ethics to replace controls based on political coercion and fears of supernatural punishment. These works continued to express faith in a view of sociocultural progress that remained simultaneously universal and Eurocentric. But, in the course of the 1850s, changing views of human origins laid the basis for a more sombre version of sociocultural evolution.

Darwinian Evolution

Until 1858, although human bones and chipped stone tools had been found associated with the bones of extinct animals, such as mammoths and woolly rhinoceroses, in caves and geological deposits in various parts of Western Europe, geologists and antiquaries had been unable to demonstrate that human beings had existed prior to the biblical date of Creation. In each case it was possible that, as a result of the resorting of sediments, human bones or tools had become mixed with older material relatively recently (Grayson, 1983: 107). In 1858 William Pengelly conducted carefully controlled excavations in Brixham Cave, near Torquay in southern England. There, chipped stone tools and fossil animal bones were found beneath a layer of stalagmitic drip 7.5 cm thick. The slow formation of such deposits suggested that the material underneath must be older than the traditional date for the creation of the world. In 1859 a team of British archaeologists and geologists examined Jacques Boucher de Crèvecoeur de Perthe's and Marcel-Jérôme Rigollot's discoveries of stone hand-axes and extinct animal bones deep in the stratified gravels of the Somme Valley in northern France. By applying the principles of the new uniformitarian geology, they were able to demonstrate that these deposits must have been formed many thousands of years ago. It was now clear that human beings had co-existed with extinct mammals at a time that was far removed from the present in terms of calendar years. This new view of the antiquity of human beings won what amounted to official approval from the greatest geologist of the period when Charles Lyell accepted it in his book *The Geological Evidences of the Antiquity of Man* (1863).

Later in 1859 Charles Darwin published the first edition of *On the Origin of Species*, which led to the concept of biological evolution becoming widely accepted in scientific circles. The key to Darwin's success was providing an acceptable explanation of how natural selection occurred that accounted both for the range of variation found in modern species and for the changes that were observed in the palaeontological record. Darwin proposed that, because all creatures vary individually and tend to increase in numbers faster than their food supplies, those individuals who are best adapted to the environment in which they live are the ones most likely to reproduce and hence pass their traits on to the next generation. The culling, generation by generation, of those individuals least able to compete in the struggle for life gradually alters the biological nature of all but the most perfectly adapted species. In 1862, Karl Marx described Darwin's natural selection as discerning 'anew amongst beasts and plants [Darwin's] English society with its division of labour, competition, elucidation of new markets, "discoveries" and the Malthusian "struggle for existence"' (cited in Sanderson, 1990: 71). Marx perceived that Darwin's ideas were acceptable because they accorded with what nineteenth-century liberals held to be common sense in the realm of human behaviour in the 1850s. They were also scientifically viable.

Yet at least part of the success of Darwin's theory resulted from the ambiguity with which he presented his ideas. In his treatment of specific cases, Darwin generally portrayed biological evolution as a fortuitous and directionless process. Unlike earlier views of sociocultural or biological evolution, he did not conceive of biological evolution either as a ladder or a tree with a central trunk leading to humanity, but as a tree of which the branch leading to human beings was only one among many. Such a presentation reflected Darwin's conviction that biological evolution had no specific goal. Many of Darwin's followers were happy enough to replace the 'anthropomorphism' of traditional theology with 'the passionless impersonality of the unknown and unknowable' (Desmond, 1994: 319). But even Thomas Huxley, and at times Darwin himself, were unwilling to embrace the chilling vision of a universe that had no meaning and no purpose. For the vast majority of Victorian intellectuals, the abandonment of traditional religious beliefs made finding some inherent meaning in the universe itself, in the form of a deistic view of evolution, all the more important.

There was a strong urge to view evolution as a progressive process, which created order and beauty out of chaos. Human beings were located on the cutting edge, if they were not the ultimate goal, of this process. This teleological view read a moral purpose into the cosmos itself and aligned Darwinism with the teleological sociocultural evolution of the Enlightenment and the evolutionist philosophy of Spencer. As Gellner (1988: 144) colourfully phrased it, it turned evolution into a God-surrogate. Sociocultural evolution was widely regarded as a continuation of biological evolution, and the ultimate goal of both as being the creation of earthly perfection. This provided nineteenth-century thinkers with a highly valued, albeit erroneous, sense of there being a preordained order and purpose to human history long after they had abandoned belief in a God who intervened directly in human affairs. While avoiding mention of human origins, Darwin combined his own selectionist view of evolution with an implicit endorsement of Spencerian teleology when he concluded the *Origin* with these words: 'Thus, from the war of nature, from famine and death, the most exalted object which we are capable of conceiving, namely, the production of the higher animals, directly follows. There is grandeur in this view of life, with its several powers, having been originally breathed [*in later editions he added* by the Creator] into a few forms or into one; and that, whilst this planet has gone cycling on according to the fixed law of gravity, from so simple a beginning endless forms most beautiful and most wonderful have been, and are being, evolved' (Darwin, 1859).

For Victorian readers, these words provided needed reassurance that the disruptions, exploitation, disease, and suffering brought about by the industrial revolution, however horrible they might be, served a higher purpose. Factory-owners and investors were not simply exploiting the working classes and indigenous peoples around the world for the financial profit they derived from them; they also convinced themselves that they were striving to create a better world for future generations of humanity (Burrow, 1966: 271–2). This adapted to the intellectual climate of the 1850s the beliefs of Benthamite utilitarians and Adam Smith *laissez-faire* economists that free enterprise, despite its emphasis on individual competition, by its very nature (through the operation of an 'invisible hand') promoted a natural harmony of interaction from which ultimately everybody would benefit (Bowler, 1984: 96). It has been suggested that Spencerian sociocultural evolution and Darwinian biological evolution were: (1) substitutes for a traditional religion that was no longer able to explain a changing social reality; (2) attempts to battle against the psychological confusion and intellectual anarchy resulting from rapid social change; or (3) a dominant group's efforts to present their exploitation of others as a form of altruism. They were in fact all of the above.

Darwin also accepted that the Lamarckian belief, that morphological changes in organs brought about by use and disuse were biologically heritable, played a lesser role in biological evolution alongside natural selection. As challenges to a very long evolutionary sequence increased in the late nineteenth century, as a result of Lord Kelvin's anti-evolutionary and, as it turned out, erroneous theory about the limited time the sun could go on producing energy, Darwin increasingly relied on Lamarckian mechanisms to increase the rate of biological evolution. This version of biological evolution encouraged a more limited teleology, which accorded significance to the goals of individual evolving organisms, to persist alongside the more grandiose cosmic teleology of Spencerian evolutionism. This too made Darwinian evolution more congenial to a Victorian audience.

In the 1860s and 1870s, conservatives who opposed all forms of evolution once again resorted to degenerationism to provide an alternative explanation of cultural differences. The leaders of this movement were Richard Whately, the Anglican archbishop of Dublin, and the Duke of Argyll (Burrow, 1966: 274). But the appeal of degenerationism was now limited to fundamentalist religious circles. Even the geologist John William Dawson's (1888) highly pertinent questioning of the adequacy of the evidence then used to support biological and sociocultural evolution tended to be ignored. This suggests that the form of evolutionism that existed in the mid-nineteenth century was fulfilling a significant social need among the middle class and the intelligentsia. From this time onward degenerationism became marginalized from the intellectual mainstream. Today it survives in a full-blown form only at the level of popular culture, among Christian fundamentalists and in speculations by Erich von Däniken (1970) and others that the evolution of human beings and of civilization results from the manipulation of human evolution by extraterrestrial visitors.

Darwinism and Sociocultural Evolution

While Darwin carefully avoided discussing human origins in 1859, the inescapable implication of his work, that human beings had not been created but had evolved from some apelike primate, made the antiquity and physical attributes of prehistoric human beings and their immediate predecessors an issue that had to be studied empirically. An evolutionary perspective also posed major questions about how the human mind and the biological basis of human behaviour had changed over time.

In the 1860s there was a growing rejection of the Enlightenment doctrine of psychic unity, which maintained that human nature and wants were essentially unchanging and hence could serve as axioms in deductive sciences of political economy and society (Burrow, 1966: 82). In place of the Enlightenment belief that human beings remained physically and mentally constant, while evolving ever more complex cultures, human nature now came to be viewed as having slowly evolved from that of an apelike ancestor. Human beings were increasingly accepted as being evolved apes rather than inferior versions of angels (Irvine, 1955). Behaviour that human beings shared with other animals, and which had long been regarded as embarrassing to acknowledge, now began to seem more important and human rationality less so. The more religious regarded this as encouraging a dangerous loss of reverence for what was distinctive about human beings and raised them above the animal kingdom. This made a more detailed understanding of human origins and evolution, both physical and sociocultural, the most vital part of the controversy that was raging over Darwin's theory of biological evolution.

Because it postulated that all human beings were descended from a single ancestral population, Darwinian evolution marked the triumph of monogenism over polygenism. No longer was it possible to maintain that different races represented separately created species. Yet the much longer period that Darwinian evolution allowed for modern races to have evolved from a common ancestor left open the possibility that extensive biological changes could have occurred during this process. This challenged the long-standing monogenic faith in the essential similarity of human nature. It was possible that over a long period natural selection, operating differently in various parts of the world, had produced human groups that were biologically very different from each other. Ironically, while it supported a monogenic theory of human origins, Darwinian evolution offered a plausible explanation for a polygenic view of human nature.

This view provided a new perspective for considering problems that American colonists had long encountered in dealing with aboriginal people and that the British were encountering in colonial situations in other parts of the world. Enlightenment philosophy had predicted that less evolved societies would quickly adopt more advanced forms of behaviour once they came to know about them. Yet this did not often occur, even where sustained efforts were made to educate indigenous peoples in European ways. Native people frequently rejected such efforts and maintained that their way of life was superior to that of Europeans, even while adopting some items of European material culture. By the late eighteenth century, many Euro-Americans had concluded that native Americans were intellectually and temperamentally inferior to Europeans in ways that could not be altered (Vaughan, 1982). At first these physical differences were explained polygenically as resulting from separate creations. Darwinian evolution attributed them to natural selection acting differently under different conditions. The state of cultural evolution attained by different ethnic groups was interpreted as evidence of the varying extent of their biological evolution.

The impact of Darwinism on studies of sociocultural evolution is evident in John Lubbock's *Pre-historic Times, as Illustrated by Ancient Remains, and the Manners and Customs of Modern Savages*, a work that passed through seven editions between 1865 and 1913 in both England and the United States, and in an even more extreme fashion in his more popular *The Origin of Civilisation and the*

Primitive Condition of Man (1870). Lubbock grew up as a neigh-
bour of Charles Darwin, whose home bordered on the Lubbock
family's estate in Kent. He became a lawyer and an amateur
naturalist whose research had established him as a leading authority on animal behaviour. It was as an early supporter of Darwin's
theory of evolution that Lubbock turned to the study of sociocultural evolution. By applying a selection of Darwinian principles
to understanding sociocultural evolution, Lubbock became the
first major exponent of what would come to be called social
Darwinism.

Lubbock did not believe that human beings necessarily had
invented all tools in the same order everywhere in the world and he
acknowledged that environmental differences produced cultural
variations in kind as well as in degree among different human
groups. But he was as deeply committed to the idea of parallel
evolution as any of the Enlightenment philosophers had been. His
overall view of cultural evolution as a ladder up which different
peoples climbed was inherently different from the dendritic pattern of speciation that characterized Darwinian biological
evolution.

What was new, and Darwinian, was Lubbock's insistence that as
a result of natural selection human groups had become different
from each other biologically as well as culturally and that these
biological differences influenced the capacity of these groups to
utilize culture. Modern Europeans were the product of intensive
cultural and biological evolution which provided them with a level
of intelligence and emotional control that was superior to that of all
other peoples. This made Europeans more innovative, more
adaptable, and better able to manage an advanced technology.
Technologically less advanced peoples were not only culturally but
also morally, intellectually, and emotionally more primitive than
were civilized ones. This was seen as a difference that in the long
run doomed the most primitive peoples to destruction. As Darwin
(1874: 203–4) was later to observe:

> an increase in the number of well-endowed men and an advance-
> ment in the standard of morality will certainly give an immense

advantage to one tribe over another. A tribe including many members who . . . were always ready to aid one another, and to sacrifice themselves for the common good, would be victorious over most other tribes; and this would be natural selection. At all times throughout the world tribes have supplanted other tribes; and as morality is one important element in their success, the standard of morality and the number of well-endowed men will thus everywhere tend to rise and increase.

Thus even a biological capacity for morality was the product of natural selection operating in a utilitarian fashion. This quotation also reveals that Darwin did not regard individual competition as the only factor contributing to the success of the human species. For him and many of his contemporaries such competition was counterbalanced by a strong sense of public responsibility.

Lubbock and many other Darwinians also believed that among Europeans, as a result of the differential operation of natural selection, women, the lower classes, and the criminally inclined were intellectually and emotionally inferior to middle-class males. This, it was argued, was because these groups did not have to participate so directly in the struggle for existence. The new industrial and the expanding professional middle class liked to believe that, as a result of heightened social competition in modern times, the biologically most fit had risen to the top of society. That sort of thinking was facilitated by a lack of understanding of the mechanisms of biological inheritance. Lubbock and the other social Darwinists used this ignorance to construct a single theory that purported to account both for social inequality in Western societies and for the assumed superiority of Europeans over other peoples.

Lubbock argued forcefully against the idea that cultural degeneration had played a significant role in human history, ridiculing it as an old-fashioned and discredited doctrine. He also opposed the Rousseauian view that the development of civilization had not led to increased human happiness. To reinforce an evolutionary perspective, Lubbock selectively used ethnographic evidence to portray primitive peoples as poor, wretched, and depraved. He described modern tribal groups as lacking both the technology and general knowledge needed to control nature. As a result they had to spend most of their time trying to secure enough food to survive and frequently encountered starvation. According to Lubbock, the minds of these people resembled those of European children. Their languages lacked abstract terms and they were incapable of understanding abstract concepts. Because of this, they were unable to comprehend natural processes and regarded the world around them with superstitious terror. They were alleged to be slaves to their passions and unable to control their anger. This resulted in much violence and further incapacitated these societies by making it impossible for people to follow a specific course of action for more than a short time. Lubbock maintained that primitive people were even more deficient in moral sense than was commonly believed and went to great trouble to document how specific groups regularly abused their children, murdered their aged parents, performed human sacrifices, and ate human flesh.

Like the Enlightenment philosophers of the eighteenth century, Lubbock maintained that cultural progress resulted in increasing population densities, but he argued that, left to their own devices, primitive peoples generally remained static or even declined in numbers. Lubbock drew upon the existing corpus of ethnographic data, which consisted mainly of descriptions of non-Western peoples by missionaries and travellers written since the beginning of the sixteenth century. On the whole these records presented a rather negative picture of simpler societies. Yet, in creating his own descriptions of modern tribal peoples around the world, Lubbock tended to emphasize negative observations and to ignore the positive ones. In his effort to make sociocultural evolution credible, he sought to make the simplest living societies appear as bestial and deficient as possible, by comparison with modern European ones. In this manner, Lubbock worked to discredit the degenerationist point of view. He also sought to establish the vast moral and biological gulf that separated modern aboriginal peoples from Europeans. Finally, by assuming that these peoples personified the various stages through which Europeans had progressed in prehistoric times, he sought to convince his readers how much

progress Europeans had achieved even prior to the earliest historical records. Lubbock was utilizing the same conjectural history
approach to the study of sociocultural evolution that had been
pioneered by the philosophers of the Enlightenment. Sociocultural
progress, however, under the influence of Darwinian biological
evolution, no longer represented the gradual realization of a fixed
human potential. Surviving primitive peoples were barred by their
biological inferiority from participating fully, and sometimes even
at all, in further cultural evolution.

Yet for the winners in this race cultural evolution remained as
hopeful as ever. Cultural development expanded human consciousness and produced growing material prosperity and spiritual
awareness. Lubbock believed sociocultural evolution would continue indefinitely into a future that would be characterized by ever
greater technological and moral improvement and by increasing
human comfort and happiness. *Pre-historic Times* ended with a
rousing summation of this credo:

> Even in our own time, we may hope to see some improvement; but
> the unselfish mind will find its highest gratification in the belief that,
> whatever may be the case with ourselves, our descendants will
> understand many things which are hidden from us now, will better
> appreciate the beautiful world in which we live, avoid much of that
> suffering to which we are subject, enjoy many blessings of which we
> are not yet worthy, and escape many of those temptations which we
> deplore, but cannot wholly resist. (Lubbock, 1869: 591)

Thus in Lubbock we also encounter an evolutionary theodicy.
The sufferings, tribulations, and strivings of the present were
justified by the role they played in helping to create a better future
for humanity. This 'deification of history' (Gellner, 1988: 141)
extended to sociocultural evolution the mission of assuring a
collective salvation for human beings that replaced or complemented the individual salvation offered by Christianity. Most
nineteenth-century sociocultural evolutionists, including Marx
and Engels, embraced the ancient Christian view of redemption
through suffering. Thus the sufferings of the working class were
justified as a necessary transitional stage in the creation of a perfect

society and hence as a price that was well worth these people paying in order to benefit future generations. This version of sociocultural evolution, like Darwinian biological evolution, reassured the insecure middle class by grounding the social order, and conventional morality, in the very fabric of the cosmos and offering a long-term justification for what it was suggested were merely the transient darker consequences of the industrial revolution.

While Lubbock is now routinely and justly denounced as a racist, many people have maintained that Darwin did not share such views. Darwin actively opposed mistreatment of other races. Yet in *The Voyage of the Beagle* he described the Fuegians as starving, dirty, ill-clad, and warlike people who would kill and eat their own elderly women before they would devour their hunting dogs. A father was reported to have thrown his infant son to his death as punishment for dropping a basket containing sea-eggs (Darwin, [1839] 1959: 205). Darwin's account of the Fuegians reads much like Lubbock's later descriptions of all hunter-gatherer peoples. Darwin was also astonished by the rapidity with which three young Fuegians who had been educated in England reverted to their traditional lifestyle.

After Darwin formulated his theory of natural selection he decided that, until palaeontological evidence of human origins was discovered, the best case for human evolution could be made if the most primitive human groups could be shown to be behaviourally as little different as possible from the great apes. He also argued that modern cultures arranged from simplest to most complex illustrated the gradual evolution of human beings, both culturally and biologically, from an original type that was perhaps less different from the great apes than it was from the most advanced modern human societies. This proposition was vehemently opposed by Alfred Wallace, the co-formulator of the theory of natural selection. He was a naturalist who had spent considerable time living among tribal societies in South America and South-East Asia and had found these peoples to be culturally different from Europeans, but not intellectually or morally inferior to them. Wallace thus questioned the link that was being assumed between biological and sociocultural evolution (Bowler, 1984: 222–3; Eiseley, 1958). This in turn led him to question whether natural selection could account for the development of human intelligence.

Darwin interpreted Wallace's views as a betrayal of their theory of biological evolution and in defence of that theory he and his followers became more intransigent in their assertion of a developmental continuity that was marked by the great apes, hunter-gatherers, and civilized human beings. Lacking adequate palaeontological evidence, neither Darwin nor Wallace was aware of the great time depth that separated the great apes from an independent hominid line. Nor were they aware of the relatively recent dates at which some hunter-gatherer cultures had evolved into food-producing ones. Through most of the intervening period, while they were living in Africa and then expanding throughout the Old World, hominids occupied a much smaller part of the earth's surface than they have during the last 12,000 years. All of them lived in small groups, and shared a similar food-gathering and scavenging way of life followed by one based on food-gathering and big-game hunting. Thus there was ample time for humans to evolve from a non-human primate ancestor. Yet throughout most of that time, because all humans lived in small groups, they were subject to similar selective forces at least as far as their behaviour was concerned. Hence natural selection could have produced human beings who were essentially alike in temperament and intelligence, as Wallace believed all modern humans are (Carrithers, 1992). In the absence of detailed information about human evolution, both Darwin and Wallace were forced to adopt positions that have turned out to be curious mixtures of correct and incorrect ideas. Darwin's belief that natural selection was the main driving force behind human evolution is widely accepted today, but it is generally agreed that Wallace was correct and Darwin wrong in their understanding of modern human nature.

Lubbock and other Darwinian evolutionists believed that the development of industrial societies was bringing about cultural and biological changes that would improve the human beings who lived in such societies and ultimately produce an earthly paradise. From a Darwinian perspective, biological improvement came about as a result both of natural selection occurring in ever more demanding social contexts and of physical and mental qualities that were being acquired in more developed contexts being inherited biologically in a Lamarckian fashion. By offering evidence that such progress was a continuation of what had been occurring at an accelerating rate throughout the entire course of human history, prehistoric archaeologists were reinforcing the confidence of the British middle class and strengthening their pride in the role they were playing in promoting human evolution.

Defending Privilege

The period from the 1850s into the 1870s represented the peak of political power and self-confidence of the British middle class. Yet the concerns of the new exponents of sociocultural evolution were markedly different from those of the French philosophers who had advocated more power for the middle class a century earlier. Enlightenment philosophers had promoted economic and political changes and a general outlook that would improve their own, as yet insecure, position. Still political outsiders, they hopefully stressed the inevitability and universality of change and described it as something from which all humanity inevitably must benefit. Once in control, however, the middle class became more concerned about protecting their political power than about extending their privileges to others, and soon they began to view a broader sharing of the economic benefits of the industrial revolution as a threat to their own interests. Nineteenth-century evolutionists tried to defend the gains of the middle class by defining natural limits to those who might reasonably expect to share in them. This shift in emphasis was reflected in theories of cultural evolution. Within the context of Lubbock's work, and social Darwinism generally, not all human groups were thought able to share in progress. The most primitive peoples, who had managed to survive in remote portions of the globe, were doomed to perish as a result of the spread of civilization, since no amount of education could compensate for the thousands of years when natural selection had failed to adapt them biologically for more complex and orderly ways of life. Nor was their replacement by more evolved peoples to be seriously regretted, except as it involved the loss of living examples of what European and other civilized peoples had been like in the remote past. Social Darwinists argued that the destruction of simpler peoples made room for the expansion of biologically more capable human groups.

This view justified European colonization and the establishment of political and economic control abroad on the ground that it promoted the progress and improvement of the human species as a whole. It also absolved European settlers of moral responsibility for the rapid decline of aboriginal peoples in North America, Australia, and the Pacific islands. It was argued that these populations were declining, not because of what colonists were doing to them, but because over many thousands of years natural selection had failed to equip them to survive as civilization expanded. Imposing subordinate status on native peoples and depriving them of their resource bases were made to appear less as political acts than as the unintentional consequences of their limited innate abilities. Whether dealing with gender issues, class, or criminality in Western societies, or with native people being caught up in the expanding European world-system, social Darwinism transferred responsibility for human inequality from the political to the natural realm by attributing it to biological differences that could be altered only very slowly, if at all. It was often asserted that, because physical changes were required, primitive peoples could not be civilized in a shorter period than it had taken Europeans to evolve to their present state (Bieder, 1986: 194–246). The new racialized evolution represented a major break with the Enlightenment's belief in psychic unity, which had expressed its hopes for the future by postulating progress in which all human beings could participate.

Racial explanations of behavioural differences were not new. Already during the eighteenth century the alleged inferiority of blacks had been used to justify the African slave trade, and the inherent primitiveness of the American Indians to explain their refusal to abandon their traditional ways of life and make room for European settlers. Hard racism of this sort was usually formulated in terms of polygenism. In the early nineteenth century, romantic nationalists had begun attributing the supposedly different behaviour of various European ethnic groups to biological differences. Social Darwinism, however, was the first theory to combine an interest in sociocultural evolution with a concern with the limitations that biological differences could impose on sociocultural development. Prior to this time, a primitive culture was thought to reflect a lack of knowledge or, at most, a temporary and reversible environmental limitation on the physical constitutions of different human groups. Now primitive cultures were interpreted as a measure of the unalterable biological inferiority of groups. At most only a few exceptional members of such groups could make the transition to a more civilized style of life. This was a transition that often involved intermarriage as well as re-education, and even exceptional individuals were likely to be absorbed only into the lower levels of the dominant society. Most people who were seen as standing on the lowest levels of the evolutionary ladder were thought bound to die out as more advanced peoples increased in numbers and pre-empted their territories. Darwinian natural selection helped to reassure a dominant middle class that their success reflected their innate natural superiority and that they would never be called on to share their privileges with either all their countrymen or the whole of humanity.

In the decades that followed the publication of *On the Origin of Species*, an evolutionary approach dominated the study of sociocultural phenomena. In France the geologist and socialist politician, Gabriel de Mortillet, was able to use stratigraphic associations to divide the Palaeolithic period into a series of *époques* or phases, each characterized by a distinctive assemblage of artefact types. These *époques* demonstrated that stone-tool technologies had become progressively more refined over time and the array of tool types more complex. Mortillet interpreted this as evidence that cultural evolution had occurred even in Palaeolithic times and assumed that, because the French sequence represented a logical developmental process, it would prove to be a universal one. Unfortunately, human skeletal material was found associated only with the later part of this sequence. Anatomically modern forms of human beings were linked with what would now be called the Upper Palaeolithic and the large-brained Neanderthal forms with the Middle Palaeolithic. This did not provide the anatomical evidence of the transition from ape to human for which evolutionary archaeologists were looking. Such forms were not discovered until much older periods of human evolution were discovered in Asia and Africa.

Evolutionary assumptions also led Palaeolithic archaeologists to misinterpret their finds. Overly rigid applications of ideas about what constituted progress caused many archaeologists to reject the authenticity of Western European cave paintings on the grounds that they were too naturalistic and sophisticated to have been produced at an early stage in human development. This mistake was only corrected when bone carvings and cave paintings were found in contexts that clearly dated to the Upper Palaeolithic period. Even then, however, European cave art was interpreted in terms of the hunting magic and totemism associated with Australian aborigines. If early cave art could not be shown to be technically and artistically primitive, then it was felt necessary to demonstrate that it had served a primitive religious function (Ucko and Rosenfeld, 1967).

While archaeology provided concrete evidence of technological evolution, the range of human behaviour that it could elucidate was insufficient to satisfy most anthropologists and other social scientists. Even archaeologists looked to primitive living societies to reveal what social life and religious beliefs might have been like in Acheulean or Magdalenian times. Among ethnologists there was a great revival in 'conjectural history', as they sought, by comparing modern societies at different levels of development, to discover the stages through which European societies had evolved in prehistoric times. At first ethnologists were interested mainly in the evolution of social institutions. Johann Bachofen (1861) argued that all societies had matrilinear beginnings. He regarded this as an idyllic state, although most ethnologists assumed that, if the earliest societies had been matrilinear, matrilineality must be a primitive and hence unsatisfactory institution. John McLennan (1865) described social organization as developing from polyandry through polygyny to monogamy. The lawyer Henry Maine (1861) used historical documentation to trace the growth of legal systems from simple forms based on status to more complex ones based on contract. While Maine's formulation was presented in an evolutionary format, he relied on examples that were historically related, and not on broad cross-cultural comparisons, and maintained that he was using the comparative method that philologists had used to trace relations among the Indo-European languages as a model for all his work (Burrow, 1966: 149).

More general evolutionary theorists included Edward Tylor, who was interested mainly in the evolution of knowledge and belief systems. He pictured early humans as primitive philosophers, applying their powers of reason to explain events that were beyond their control in the human and natural realms, though in their scientific ignorance they frequently produced erroneous explanations. Tylor concluded that the earliest form of religion was animism, a belief in spiritual beings that had been arrived at by attempts to differentiate living bodies from corpses and to account for the apparent separation of mind from body experienced in dreams. Tylor sought to use anthropology to bring about moral and intellectual progress by weeding out irrational elements in modern society. He tried to do this by identifying survivals: aspects of older systems of thought that persisted unchallenged at later stages. He identified survivals by means of a historical approach, and interpreted them as evidence of cultural continuity and a major proof that cultural evolution had taken place (Burrow, 1966: 254–5). Tylor's work was carried on by James Frazer, the author of *The Golden Bough* (1911–15), who sought to delineate an evolutionary sequence of magical, religious, and scientific thought.

Lewis Henry Morgan was an American lawyer who progressed from studies of the evolution of kinship systems to a far more comprehensive study of the co-evolution of technologies, social institutions, and intellectual culture. Morgan (1877) saw human societies as evolving through various subphases of savagery, barbarism, and civilization that were distinguished by specific technological innovations. The criteria for these substages differed somewhat for the Old and New Worlds because of the long-recognized failure of most native Americans to elaborate systems of food production involving domestic animals. While Morgan is frequently cited as according primary or determining importance to economic factors, he tended to regard the various parts of a sociocultural system as interrelated and mutually influential. More than any other nineteenth-century evolutionary anthropologist, he viewed societies as functionally integrated systems. Like other parallel evolutionists, he supported a unilinear view of cultural development, which postulated that no culture could reach a higher level without first passing through all the lower ones. Yet he also believed that by abolishing hereditary privileges, moving to manhood suffrage, and introducing universal education, American society, which he believed was the most advanced in the world, was reviving at the highest level of development the equality of an earlier clan society. In Morgan's words: 'Democracy in government, brotherhood in society, equality in rights and privileges, and universal education, foreshadow the next higher plane of society to which experience, intelligence and knowledge are steadily tending. It will be a revival, in a higher form, of the liberty, equality and fraternity of the ancient gentes [clans]' (Morgan, 1877: 552). This vision appealed greatly to Marx and Engels.

Another influential exponent of sociocultural evolution was the American sociologist and economist William G. Sumner. Inspired by Spencer, he expounded the belief that a combination of economic *laissez-faire*, untrammelled personal liberty, and unimpeded competition for property and social status constituted the ideal toward which all societies should strive. Considering poverty to result entirely from innate inferiority, he opposed social welfare as being contrary to the best interests of society and the individual. He believed that customs originated in instinctive responses to hunger, sex, vanity, and fear and emphasized the frequent irrationality of folkways as well as their resistance to reform (Sumner, 1906).

Concepts of sociocultural evolution pervaded every type of study of human behaviour in the second half of the nineteenth century, including archaeology, anthropology, political economy, sociology, history, and psychology. Yet, while attempting to identify with and share in the prestige of Darwinian evolution, sociocultural evolution remained distinct in many ways and continued to have important features in common with the sociocultural evolution of the Enlightenment. In place of Darwin's tree of life, with its branches spreading fortuitously into a panoply of different environments, sociocultural evolution continued to be viewed as a single ladder or series of adjacent ladders that led upward in the same direction. Living 'savages' were viewed as Stone Age relics 'trapped out of time in the colonial backwaters' and providing 'snapshots of our primitive past'. They were 'children' to be tended by white 'adults' (Desmond, 1994: 342). The only exception to this view was found in linguistics, where Max Müller treated languages as developing in the form of branching trees with no privileged trunk and maintained that each language exemplified the common underlying potential of languages as a whole (Bowler, 1989: 68).

Culture was viewed as humanity's specific form of adaptation and believed to develop along similar lines, although not at the same rate or to the same degree, in different parts of the world. Because of this, it was not regarded as important to understand how specific cultures had evolved. Some sociocultural evolutionists, such as Tylor, might seek to discover how specific aspects of human behaviour, such as kinship systems or kinds of religious beliefs, had evolved, while others, such as Morgan, might wish to examine the development of societies more holistically. It was assumed that, since all societies developed along similar lines, an understanding of culture in general would permit social scientists to understand at least in principle how specific cultures had developed.

Repudiating the Enlightenment?

Where late nineteenth-century sociocultural evolution differed most obviously from that of the Enlightenment was in its racism. After 1865, under the influence of Darwin, Huxley, and Lubbock, most anthropologists tended to assume that less developed peoples were mentally, emotionally, and physically inferior to Europeans. Even Tylor, who initially had believed in the uniformity of human nature, eventually embraced a biological explanation of cultural differences (Bowler, 1992: 726–7). Yet racial explanations played a more prominent role in discussions of colonialism and in formulating policies for dealing with aboriginal peoples than they did in general analyses of cultural change. Racial explanations may have been used to account for why different groups progressed at different rates, supplementing or replacing the environmental explanations of an earlier era. Nevertheless, most evolutionists continued to believe that culture developed along similar paths independently of what was happening elsewhere. This was possible only if all human beings at the same level of development tended to think in essentially the same way and human thought naturally developed along similar lines. Evolutionists thus continued to believe that the minds and behaviour of one group of human beings were essentially similar to those of another; the main difference being the greater or lesser extent to which they were developed in relation to a unilinear standard. What evolutionists no longer believed was that differences in level could be overcome quickly, if at all.

Morgan and Tylor saw similar human institutions and beliefs developing everywhere from a few germs of thought as the result of general likenesses in human nature and in the circumstances of human life. The German ethnologist, Adolf Bastian, who in 1851 began a series of voyages around the world to collect material for the Royal Museum of Ethnology in Berlin, was impressed by the cultural similarities that he encountered in widely separated regions. He argued that, as a result of universally shared 'elementary ideas' (*Elementargedanke*), peoples facing similar problems will, within the constraints imposed by their environments, tend to develop similar solutions. Ethnologists thus continued to a considerable degree to analyse their data in terms of the doctrine of psychic unity, which they only abandoned in order to explain differing rates of cultural evolution.

It has frequently been maintained that, while nineteenth-century parallel evolutionists treated sociocultural evolution as part of the natural order of things, they did not conceptualize any mechanism for explaining how it happened (Sanderson, 1990: 20). It is true that sociocultural evolutionists, such as Morgan, sometimes spoke of institutions logically unfolding from initial conditions, which seems to imply that change is immanent and does not require explanation. Yet, like the Enlightenment philosophers, evolutionists of the nineteenth century saw the main driving force behind change being individual humans applying their powers of reason to enhance the quality of life by controlling nature more effectively and to improve knowledge, behaviour, social institutions, and living conditions generally. Given these motivations, evolution was interpreted as the progressive elaboration of a few basic germs of thought. While material factors, such as environmental changes and increasing population, had long been viewed as external forces that encouraged cultural development, the basic understanding of sociocultural evolution remained rooted in the idealist approach of the Enlightenment (Sanderson, 1990: 21–6). It was not that nineteenth-century evolutionists lacked a theory of change, but rather that there was a vast amount of continuity between what was believed by sociocultural evolutionists in the eighteenth and nineteenth centuries despite an increasingly racist view of human behaviour. While psychic unity had been abandoned when it came to explaining different rates of sociocultural evolution, it continued to provide an explanation of change itself. These contradictions reflected the moral and conceptual dilemmas that Victorian society faced in trying to understand the social and economic changes that were occurring at mid-century. Differences of a similar magnitude occurred in biological evolution as the result of a morally-driven desire to accommodate the Lamarckian principle of the inheritance of acquired characteristics within natural selection.

Methodological Advances

Today the sociocultural evolution of the second half of the nineteenth century is often equated with Mortillet's belief that stone tools everywhere would evolve through the same unilinear sequence that had been discovered in France. This has long been demonstrated not to have been the case (Ikawa-Smith, 1978; Movius, 1948). It is obvious that, despite their adherence to unilinear evolution as a general concept, most Victorian scholars did not attempt to force their data into a strictly unilinear framework. Lubbock, for example, doubted that specific types of tools had been invented in the same order everywhere.

Already in the eighteenth century evolutionists were aware that geographical factors complicated the picture. Eskimos could not be expected to have become farmers and the development of food production in the Old World, where domesticable species of animals were abundant, followed a different course from that in the Americas, where such animals were not. Evolutionists remained aware of these and other environmental problems in the nineteenth century. Many European historians maintained that harsher northern climates were responsible for the later development of civilization in such regions, but that eventually their demands stimulated greater progress than did more southerly ones. The English historian Henry T. Buckle (1857) argued that, because of the easy conditions under which tropical civilizations developed, they tended towards despotism, pomp, and sensuous display rather than promoting the mental and moral progress of the masses as did cultural development in more temperate climates. The idea of the superiority of northern climates for cultural development was to play a major role in anthropological thinking and nationalist rhetoric throughout the rest of the nineteenth century (Berger, 1970; Keen, 1971: 443–4). The historian Thomas Macaulay (1858) also argued that, because Britain was an island and therefore had no need of a standing army, it had leapfrogged from being a feudal monarchy to being a modern liberal state without passing through a stage of absolutism as its continental neighbours had done.

Evolutionists were also aware of the impact of diffusion. In 1889, on the basis of statistical correlations between marriage customs and the types of subsistence economies found in various modern societies around the world, Tylor argued that matrilocal residence appeared to have preceded patrilocal residence because today it was associated with more primitive economies. Francis Galton, who attended Tylor's talk, questioned his conclusion by asking Tylor if he had any means of knowing in how many cases these traits had been spread from one society to another by diffusion or migration and in how many they had been independently invented. Unable to answer this question, Tylor abandoned this approach, which he had hoped would refine the techniques of conjectural history. Sociocultural evolutionists had never ruled out diffusion and migration as processes that helped to bring about cultural change, hence Tylor acknowledged that his approach was vulnerable to this objection. Nevertheless, Tylor had pioneered a form of cross-cultural investigation that would become important for studying cultural development in the second half of the twentieth century (Murdock, 1949).

The most valuable empirical contribution to the study of sociocultural evolution was made by General Augustus Lane-Fox Pitt-Rivers. In the 1850s he had undertaken a detailed study of the history of firearms while serving on a committee to select a new rifle for the British army (Pitt-Rivers, 1858). This study revealed that, while guns had become more complex and efficient over time, they had not done so in a unilinear fashion. The development of firearms had involved a vast number of innovations, few of which proved to be successful. Hence the development of the rifle was characterized by many dead ends. Throughout the 1860s, Pitt-Rivers built up a vast ethnographic collection and wrote about primitive warfare, navigation, and principles of classification. He stressed that innovation was more often the result of accident than it was of premeditation or design. When it came to working out histories of innovation, it was necessary to examine specific sequences rather than simply delineate overall trends. The latter, to the extent that they had any historical validity, were a reflection of what had worked rather than a record of what had been invented (Pitt-Rivers, 1906). By studying rather than merely speculating about the nature of cultural evolution, Pitt-Rivers realized that, however much humans might be trying to improve material culture, specific innovations were rarely made on the basis of a clear understanding of even their short-term consequences. Hence cultural evolution was not as teleological or as unilinear as it appeared to be in retrospect. Of all the Victorian sociocultural evolutionists, Pitt-Rivers contributed the most to understanding the process of cultural change.

6
Revolution and Solidarity

By the end of the nineteenth century biological and sociocultural evolution generally were accepted as historical facts by the scientific community. Evolutionism had won its struggle against creationism, as far as such a victory is ever possible. A growing archaeological database confirmed that everywhere a long period of hunting and gathering had preceded the development of agriculture and civilization. Pitt-Rivers was studying how cultural change actually occurred and it seemed likely that more research of this sort would produce a clearer understanding of evolutionary mechanisms. These accomplishments suggested that sociocultural evolution was going to have a bright future. Already, however, it was under attack and in the West was about to pass into an eclipse that would last for over fifty years.

Social Challenges

By the 1880s growing social and economic strains were encouraging the dichotomization and radicalization of political views in the Western European heartland of evolutionary thought. The problems of the industrial revolution were becoming increasingly evident, especially in Britain where it had been going on the longest. These took the form of slums, environmental pollution, economic crises that resulted in high levels of unemployment, and growing international competition for foreign markets. It became increasingly difficult to dismiss these problems as being part of a transitional phase in human history that would eventually produce a better life for all.

Influenced by such considerations, a younger generation of intellectuals, led by the art critic John Ruskin and the designer William Morris, turned against the idea of progress, romantically preferring what they saw as the comforting religious faith, mutual support, and aesthetic values of an earlier age to the selfishness, materialism, and ugliness of industrial society (Bowler, 1989: 43–4). Industrialism, which in the 1850s had been a source of pride and hope for the future, was now held responsible for social chaos, squalor, and spiritual depravity. In the political realm, anarchists reasserted the Rousseauian claim that human beings were essentially good and mutually supporting and denounced industrial societies for undermining communitarian values that had been preserved as late as the medieval period (Kropotkin, 1902).

The First World War further indicated that there was no necessary connection between technological development and moral improvement (Burrow, 1966: 277). In his influential study *The Decline of the West* ([1918–22] 1926–8), the German philosopher Oswald Spengler proclaimed that Western civilization had already passed through its creative phase and that its future could only be one of irreversible decline, without hope that any of its significant achievements could be passed along to a future civilization.

By the 1880s the Western European middle class was also facing a growing political challenge. Marxists were arguing that the problems confronting the working class could be resolved only if workers were to seize political control from the middle class and landowning aristocrats who currently monopolized power. Since the ruling classes were unlikely to relinquish their privileged position voluntarily, Marxists maintained that the working class would sooner or later have to seize power by revolutionary means. European Marxists were working to enhance the international solidarity of the working class, so that workers in different countries might co-operate in toppling their class enemies. Marxists also stressed the primacy of class interests as a factor shaping history and minimized the significance of biological and cultural differences among national and ethnic groups. Because they were anxious to bring about major social and political changes, Marxists continued to adhere to Enlightenment philosophy in stressing the flexibility of human behaviour and the ability of human beings to use reason to bring about changes that were in their personal economic and political interest and that of society as a whole. At the same time, in countries that had a reasonably broad electoral franchise and democratic traditions, labour movements began to develop that sought to wrest power from the ruling classes at the ballot box. Some of these movements succeeded in educating and organizing large numbers of workers and developed into mass political movements. By 1913 the German Social Democratic Party had over 1 million members.

Seeking Social Solidarity

In order to counteract the political threat that was being posed by the working class, historians, political economists, and anthropologists in the countries of Central and Western Europe began to lay increasing emphasis on the importance of national unity. This was a broadly based and apparently spontaneous intellectual reaction that, unlike Marxism, had no obvious founder or outstanding leader. In *Physics and Politics* (1872) the British political economist Walter Bagehot had argued that strong nations dominate their neighbours and contribute to the development of civilization, while over time weak nations, whatever their merits, are subjugated or eliminated. Strength is achieved by the subordination of the individual to society. Bagehot maintained that it was the duty of church and state to strengthen the nation by suppressing any freedom of thought that threatened its unity. The evolutionary biologist Ernst Haeckel advocated a united Germany and supported a liberal democratic form of government. He too, however, argued the need to turn the German people into a purposeful social entity and lauded the Prussian army as the protector of the newly-created German state (Bowler, 1984: 272–3).

In their efforts to counteract the threat of growing class warfare, historians and anthropologists laid still more emphasis on biological factors as determinants of human behaviour. They argued that the French, Germans, and English were biologically different from each other and that their allegedly distinctive behaviour was determined not so much by economic, political, and cultural factors as by ancient and immutable racial differences. To promote national unity it was argued that within nation states, everyone, or at least all true members of the ethnic group on which the nation was founded, regardless of their social class, were united by a common biological heritage that constituted the strongest and most stable of all human bonds. Thus an English worker had more in common with his English employer than he could ever hope to have with a French or German worker. The Marxist call for the workers of the world to unite was countered by claims that workers in different countries could never understand each other's behaviour, let alone agree on common policies. Much ingenuity was expended in arguing the ethnic unity of nation states and establishing that ethnic differences corresponded with modern political boundaries. As economic and political conflicts within nation states grew more acute and competition between them increased, there was, as there is today (Harvey, 1995), a growing appeal to nationalism as a basis for maintaining unity.

Increasing emphasis was also placed on the social Darwinist belief that within modern nation states class and status largely reflected the relative biological fitness of individuals. It was assumed that, since the advent of free enterprise, the most talented, creative, and energetic individuals had risen to the top and passed their superior biological traits on to their descendants. It followed that any attack on the existing social order was a counterproductive effort to undermine an immutable biological one. Francis Galton expounded the idea that mental as well as physical characteristics were inherited in his *Hereditary Genius* (1869) and devoted the rest of his life to improving the physical and mental condition of the human species by means of selective parenthood. Even Charles Darwin, who claimed previously to have believed that 'excepting fools, men did not differ much in intellect, only in zeal and hard work', professed to have been led to view intelligence as far more innate and biologically determined as a result of his cousin's research (Desmond and Moore, 1992: 572). Galton coined the term *eugenics* for his movement in 1883. One of the main concerns of eugenics was that the working class was reproducing faster than the supposedly biologically fitter middle and upper classes. In the early twentieth century social workers who subscribed to the ideas of the eugenics movement sought to counteract this process by advocating the sterilization of those they judged to be feeble-minded, insane, or suffering from severe and inheritable physical disabilities. It is difficult to determine how far eugenics doctrines were accepted by the working class. Eugenics was, however, a means of coercing the lower orders, while at the same time it helped to reassure the dominant middle class that their privileged position in the scheme of things was natural rather than accidental.

Disillusionment about the value of technological progress, combined with a growing belief that human behaviour was biologically determined, produced increasing scepticism about human creativity. Social analysts began to maintain that change was contrary to human nature and psychologically harmful to human beings. The sociologist Emile Durkheim (1893, 1895) argued that a healthy society was one characterized by social solidarity and that societies experiencing rapid social change, such as his own, produced feelings of anomie and alienation. Yet he also believed that, as societies grew more complex, they ceased to be held together by mechanical solidarity, or shared beliefs, and were integrated to an ever greater degree by organic solidarity, which was the result of growing economic interdependence. Durkheim valued this new form of social cohesion because it freed individuals from the tyranny of custom and tradition.

The belief that a static condition was best suited to most, or even all, human beings was congenial to scholars who were anxious to resist change in their own society. This view became increasingly common in the late nineteenth and early twentieth centuries. It discouraged belief in the likelihood of parallel cultural development, or the same thing being invented more than once in the course of human history. This resulted in growing reliance on migration and cultural diffusion to explain cultural change. It also encouraged increasing interest in the idiosyncratic features associated with particular ethnic groups rather than the universal characteristics of successive stages of cultural development. If, in the 1860s, the desire of the British middle class to monopolize the political and economic gains they had made had induced Lubbock to abandon the concept of psychic unity in favour of a biological explanation of the behaviour of different ethnic groups, the still greater political insecurity of the 1880s led a growing number of intellectuals to jettison the Enlightenment doctrine of progress and regard all human beings as naturally conservative and resistant to change.

The late nineteenth and early twentieth centuries were a time of intellectual debate and confusion. There was a widespread tendency in Western Europe and North America to reject the positivist certainties of a previous era and to turn away from science towards various forms of religious and mystical experience. Idealism and romanticism both flourished. It was also a period when spiritualism attracted much attention among the middle class. In biology a materialist belief in natural selection lost ground to Lamarckian-based forms of vitalism, which attributed the evolution of new species of plants and animals to directive forces that were inherent in life itself. The result was a patchwork of differing views of cultural change that were linked to different political commitments and varied in intensity from one country to another. Evolutionism remained strong among Marxists, but among liberal and conservative scholars opinions ranged from cautiously evolutionist to violently anti-evolutionist.

A conservative evolutionary perspective survived in the work of Albert G. Keller, who continued the Sumner tradition at Yale University, while the socialists L. T. Hobhouse, G. C. Wheeler, and M. Ginsberg (1915) continued to seek for evidence of evolutionary trends in ethnographic data. In general, however, outside the Marxist tradition an interest in sociocultural evolution withered. The rest of this chapter will examine the politically charged social theories of the late nineteenth and the first half of the twentieth centuries.

Classical Marxist Evolutionism

Marxism, in keeping with its desire to bring about radical social change, emphasized the plasticity of human behaviour, and this in turn encouraged its adherence to an evolutionary perspective. While Marxism preserved many features of the Enlightenment, such as a commitment to psychic unity and an antipathy to racism, it developed a radically different view of cultural change. Marxists rejected the idea that progress came about as a result of human beings gradually acquiring more objective knowledge, which permitted them to control nature and themselves more effectively. Instead, adopting a materialist version of Hegelian philosophy, they interpreted technological and social development as a dialectical process brought about by gradual increases in class conflict and its periodic revolutionary resolution. Overcoming complete domination by nature had led to a situation in which a succession of different classes had dominated and exploited the rest of society. It was believed that such unjust relations could be overcome only after capitalism had been overthrown and a triumphant working class had seized power. While it was acknowledged that economic productivity had increased dramatically under capitalism and created the potential for a better life for all, it was also believed that injustice and human suffering would increase until a socialist revolution had created a new society that could utilize the technology that capitalism had created to promote the common good.

Marxism differed from Enlightenment philosophy in its strongly materialist orientation. Like other nineteenth-century social theorists, Marx viewed human beings as having evolved the ability to co-operate as members of social groups to the extent that they were able to transform their relations to the natural world more than any other animal could. Yet Marx differed in identifying the organization of labour, rather than the disembedded exercise of individual creativity, as being the most important means by which humans could confront nature as one of nature's own forces. He saw the economic base as playing an important role in shaping other aspects of society, such as concepts of property, family life, political organization, law, religious beliefs, aesthetics, and scientific inventions. The central tenet of Marxism declared that it was not the consciousness of human beings that determined their existence, as idealists believed, but, on the contrary, social existence that determined their consciousness. Marx did not believe that technological change resulted from human beings using their intellects to develop more effective ways to control their environment. While new technologies bring about social and political changes, they themselves are the products of specific social contexts that influence what innovations are likely or unlikely to occur.

If social Darwinists saw individuals as the chief competing elements, and hence the units on which the sociocultural equivalent of natural selection operated, Marx attributed this role to classes competing to control the forces of production. Class conflict altered the ways in which individual human beings related to one another in social and political as well as economic terms. Marx's subordination of the individual to the class as a unit of struggle reflected the general tendency of Hegel and the German romantics to subordinate the individual to the group (Bowler, 1984: 100–1). This contrasted with the pleasure principle of British utilitarian philosophy, as expounded by Jeremy Bentham and later by John Stuart Mill, which, by postulating that individuals act in ways that increase their personal pleasure and minimize their pain, tended to reinforce methodological individualism.

Marx endeavoured to explain concrete historical events as well as to delineate general evolutionary trends in human history. He did not distinguish between history and evolution as many later scholars were to do. Marx accepted that specific passions and rational perceptions of material self-interest which were common to the vast majority of human beings played a key role in determining individual behaviour. Yet he also agreed that the behaviour of a critically placed individual could play an important role at a time of crisis. Hence his more detailed studies of social change stressed purpose and the social reproduction of reality rather than treating human behaviour as a passive response to social forces. He believed, however, that in the medium or long term history has to be understood as the collective outcome of social and economic groups seeking to preserve and enhance their wealth and power in relation to each other.

Marx also believed that every society has within itself tendencies that both promote and oppose social change. Again drawing on Hegelian philosophy, he argued that each society contains the seeds of the destruction of its present state and the embryo of a future, even more developed one. The antagonism between the conservative forces that in their own interest seek to maintain the *status quo* and the progressive forces that, likewise in their own interest, seek to bring about change constitutes the main dynamic that shapes human history. Marx did not deny that factors such as entrenched political hierarchies and powerful religious beliefs can play a major historical role, but he argued that this role was a negative one, being limited to inhibiting or preventing progressive change. Progressive change, Marx believed, could occur only when economic transformations were not overwhelmed by reactionary forces.

Marx believed that every society was the product of its own unique history and responded to economic changes in a distinctive fashion. This occurred because people, while able to judge material self-interest in terms of universal standards, nevertheless inevitably perceived reality through their own distinctive culture. Hence every culture, within the context of changing relations of production, provided the material for shaping a future society. The fact that nineteenth-century Britain was governed by a monarch, a House of Lords, and a House of Commons could not be explained by its capitalist mode of production. This arrangement was the continuation of a system that had evolved long before under feudalism. On the other hand, the roles that were played by these three entities in governing nineteenth-century Britain could not be explained by their feudal origins, but had been determined as a result of the industrial revolution enhancing the power of the middle class, which led to the increased importance of the House of Commons. Engels ([1890] in Marx and Engels, 1962, 2: 489) similarly observed that 'without making oneself ridiculous it would be a difficult thing to explain in terms of economics the existence of every small state in Germany, past and present, or the origin of the

High German consonant shifts'. Because every society is a transformation of its own antecedents, it is impossible to formulate
general laws that would explain the concrete reality of sociocultural
change in a predictive fashion.

Marx and Engels thus drew an implicit, but clear, distinction
between what would later be defined as society, or systems of social
relations, and culture, or the patterns of ideas that mediate human
behaviour. They also made progress in defining how these two
entities changed in relation to each other. Whether or not one
agrees with Marx, he propounded a theory of change that was far
more detailed and sophisticated than any previous one had been
and thus provided a superior basis for discussion and disagree-
ment.

In some of Marx's writings there is a suggestion that he believed
in multilinear evolution, at least in the short and middle term. His
concept of the Asiatic mode of production implies that he saw
sociocultural evolution following a different trajectory in China
and India from that followed in the Middle East and Europe. His
discussion of the German mode of production further suggests that
he regarded early German socioeconomic arrangements as distinctive (Hobsbawm, 1964). Yet he also believed that the ideal and
inevitable course of sociocultural evolution ran from primitive
egalitarian societies, through various forms of class societies, to the
technologically advanced, egalitarian societies of the future. On
the ground that it would be unscientific to do so, Marx refused to
speculate about the precise nature of the classless societies that
would be created after the working class had seized power. Yet he
looked to the preclass societies of his own day as evidence that
greed and exploitation were not inherent in human nature but
instead had arisen, as Rousseau had believed, in the context of
class societies and would disappear in the highly developed classless societies of the future. In his own day humanity was passing
through a phase of suffering and conflict toward a future when the
political triumph of the working class would free all human beings
from exploitation and personal alienation. While many eighteenth-
and nineteenth-century sociocultural evolutionists hoped for the
gradual development of a better future for humanity, Marx, once

again inspired by Hegelian philosophy, envisaged a historical drama that in its salvationism rivalled the Christian doctrine of the return of Christ.

Unfortunately, by presenting the outcome of history as pre-determined and Marxism itself as a scientific and predictive account of history rather than a political programme, Marxists tended to minimize the important role that moral choices as well as material self-interest played in shaping social change. This trend, which was most pronounced in popular versions of Marxism, sought to convince its followers of the infallibility and ultimate success of their movement. Yet, by stressing teleological and economic factors rather than class struggle as the key factors bringing about social change, it encouraged the view that any action which served to advance the preordained course of history was justified. The prominence accorded to such views, which did not accord with Marx's more nuanced understanding of the nature of social change, played a major role in undermining the moral integrity of the movement and justifying the crushing of internal dissent as Marxism grew politically more powerful. This was another price that Marxism paid for its Hegelian connections.

Soviet Evolutionism

A still more extreme and dogmatic version of Marxist sociocultural evolution was developed in the Soviet Union after 1928. Rooted in the writings of the pre-revolutionary Russian Marxist theoretician Georgy V. Plekhanov, and simplified for public indoctrination under Joseph Stalin, Soviet evolution stressed a highly unilinear and deterministic view of social change. Social evolution was conceptualized in terms of a succession of socioeconomic forma-tions loosely derived from Frederick Engels's *The Origin of the Family, Private Property and the State* ([1884] 1972), which in turn had been based on Marx's unpublished reinterpretation of Mor-gan's *Ancient Society* (1877). Preclass societies were seen as evolving through successive preclan, matrilinear clan, patrilinear clan, and terminal clan stages; followed by three forms of class society: slave, feudal, and capitalist; and two more forms of class-less society: socialist and communist. The latter were represented as the final stages of human development. This sequence was accorded canonical status by the Soviet government and social scientists were required to interpret their data in conformity with it, rather than using their findings to question or even refine any aspect of it.

Soviet evolutionary theory was also influenced by the ideas of the linguist Nicholai Marr, which were considered to accord with Marxist thought until Stalin denounced them as nonsensical in 1950. Marr disagreed with the generally accepted belief that languages change as the result of slow and largely unconscious phonological, lexical, and grammatical differentiation from ancestral forms. He believed that linguistic change occurred as a response to alterations in the socioeconomic organization of the societies in which their speakers live. Hence similarities among languages are indicative of the stage of evolution these societies have reached rather than of historical affinities. This scheme allowed historians and archaeologists to ignore even the most blatant linguistic evidence of ethnic movements and to interpret the record of each region from earliest times to the present as stages in the development of a single people. The archaeologist Vladislav I. Ravdonikas, for example, argued that in the Crimea Iranian-speaking Scythians had evolved into 'German'-speaking Goths (whose language was historically unrelated to the German spoken farther west), and finally into Slavs. The archaeological record of each region of the Soviet Union was thus interpreted as representing the prehistory of the same people who lived there in modern times and cultural changes were treated as being of local origin. Both diffusion and migration were stigmatized as counter-revolutionary attacks on human creativity. Marr's extreme version of parallel evolution was hailed as a rejection of the anti-evolutionist and often racist theories of cultural development that were prevalent in the West. Physical anthropological studies that sought to distinguish ethnic groups on the basis of physical differences were prohibited.

Recently it has been suggested that Marr's scheme was supported by the Soviet government as a means to justify the assimilation of ethnic minorities into a dominant Russian culture. It was assumed that the Russian language and culture were superior to all others found within the Soviet Union and therefore were the goal towards which other cultures would spontaneously evolve (Shnirelman, 1995). This may have been so as Russian nationalism intensified, beginning in the late 1930s. Before that time, however, Soviet parallel evolution's main political purpose appears to have been to promote the idea that all peoples were equally creative and hence that, when relieved of Tsarist exploitation, all the ethnic groups in the Soviet Union were equally capable of developing societies that would be Soviet in form while remaining national in content. This view corresponded with Lenin's desire to modernize as quickly as possible the exploited and underdeveloped regions of the former Russian Empire while respecting local cultures. Whatever the motivation, the result was the most extreme version of parallel evolution ever formulated; one that saw each ethnic group evolving *in situ* and with neighbouring peoples exerting practically no influence on each other. This was a construction that made little sense in terms of what is, and was, known about ethnic history and cultural diffusion in historical times. Such a theory could only flourish unchallenged in a country where questioning party directives about even such esoteric matters as prehistory could result in execution or incarceration in the Siberian gulag.

Western Culture-Historicism

In the United States John Dewey, Charles Peirce, William James, and some other late nineteenth-century social scientists accepted sociocultural evolution but, in accordance with Darwin's understanding of how biological evolution worked, denied that it moved towards a predetermined goal. Thomas Huxley likewise acknowledged that nature was without purpose and changed without reference to human values. Yet, as an ardent opponent of social Darwinism, he argued that civilization is valuable precisely because it allows human beings to progress morally by using their acquired knowledge to resist the natural selection that governs biological evolution. Such resistance took the form of protecting the sick and the weak (Huxley, [1893] 1947). These attitudes represented a reaction against the hard-line sociocultural evolutionism of the mid-nineteenth century.

Yet most Western historians and social scientists went further, and, while not denying that all modern human societies had evolved from earlier hunter-gatherer ones, preferred to abandon all evolutionary explanations of sociocultural change. They rejected the claim that change came about as a result of the same innovations being made independently in different places as they were needed. Instead they relied on migration and diffusion to explain the changes observed in the historical and archaeological record. Migrationary explanations tended to be more pessimistic about human creativity than were diffusionist ones. Diffusionists assumed that humans were better able to copy other people's ideas than to innovate. Migrationists believed that people were unable or unwilling to change their ways and therefore all cultural change had to come about as the result of immigrants with their distinctive cultures replacing or blending with established groups.

In the United States it had long been thought that aboriginal peoples were wholly uninventive and that their cultures had been static prior to the arrival of European colonists. As a result, these peoples continued to illustrate the earliest stages of human development. As the archaeological record grew better known, it became evident that changes had taken place, but these were interpreted as indicating that unchanging and often warlike peoples had shifted about on a large and thinly populated continent. It was fashionable to believe that in prehistoric eastern North America a semi-civilized Moundbuilder people had been annihilated or displaced by the savage ancestors of the modern Indians (Silverberg, 1968). In New York State an archaeological sequence was interpreted as evidence that the ancestors of the present-day Inuit, Algonquians, and Iroquois had lived there in turn. Only as it became clear that such migrations could not explain the complex patterns of change observed in the archaeological record was it

accepted that trait diffusion had played a significant role in bringing about cultural change. Yet, with rare exceptions, such as Alfred
Kidder (1924), American archaeologists were reluctant to ascribe
more cultural creativity to the ancestors of the American Indians
than was needed to explain their findings (Trigger, 1980).

At the same time that American anthropologists slowly and
grudgingly began to acknowledge the creativity of aboriginal peoples, among European intellectuals there was increasing interest in
diffusion and migration as a belief in all human creativity and
therefore in evolution declined. The German geographer Friedrich
Ratzel rejected the concept of psychic unity which Adolf Bastian
had been championing. Ratzel argued that the world was small and
social scientists must not believe that even the simplest or most
generic innovation was likely to be made more than once. He
proclaimed, in a purely self-serving and arbitrary fashion, that it
was necessary to rule out any possibility of diffusion having occurred before it could be suggested that the same type of artefact had
been invented more than once. He attempted to demonstrate that,
despite their discontinuous and widely separated occurrences,
items such as the blowpipe and barkcloth, which were found in
both the Americas and South-East Asia, were derived from a single
origin. He also argued that cultural change was capricious.
Although Ratzel recognized that material constraints sometimes
compelled related aspects of technology to develop in a particular
order, he maintained that it was impossible to predict how or under
what circumstances new inventions would be made. It was also
impossible to predict whether a particular human group would
adopt even a highly useful innovation. Yet Ratzel also sought to
show how, despite its capriciousness, diffusion created culture
areas as people borrowed more ideas from close neighbours than
they did from people living farther away.

Ratzel's theories reflected German romanticism's emphasis on
idealist explanations of human behaviour and its preference for
studying cultural specifics rather than cross-cultural uniformities.
The most significant theoretical development of Ratzel's approach
was carried out by the German physicist-turned-ethnologist Franz
Boas. Boas opposed the concept of sociocultural evolution and

argued in a Herderian fashion that every culture was a unique entity that had to be understood and appreciated on its own terms. Like Ratzel, Boas viewed cultural change as a fortuitous process that came about as a result of diffusion, miscopying of traditional ways of doing things, and chance recombinations of existing ideas. Yet increasingly Boas and his students also came to view cultures as assemblages that exhibited internal patterning and coherence. Each culture had to provide for certain functional prerequisites, such as food and shelter, if its members were to survive and reproduce themselves in sufficient numbers to sustain their way of life. Environmental factors might limit, though they could never determine, how these functional prerequisites were satisfied. Still more importantly, individual cultures required psychological and logical consistency if they were to satisfy the emotional needs of their members. Ruth Benedict, in her extremely popular *Patterns of Culture* (1934), which distinguished between Apollonian cultures that emphasized co-operation and Dionysian ones that encouraged individual rivalry and competition, saw these patterns as being created by internal selection. This selection took the form of individual choices unconsciously being made by people in accordance with the expectations of their culture (Ingold, 1986: 191–6). As early as 1913 the Boasian anthropologist Alexander Goldenweiser had argued that, since the range of traits any one culture may be able to integrate successfully is limited, features that are different to begin with often end up being channelled along similar lines.

In addition to adopting an idealist view of culture, Boasian anthropology was built on two other key propositions. The first was cultural relativism. Boas maintained that no standard could be used to compare the relative degree of development or intrinsic worth of different cultures. It was particularly unacceptable to judge cultures by the extent to which they resembled the anthropologist's own. Instead they had to be evaluated in terms of the meaning they had for their own people. The second proposition was historical particularism. Each culture had to be understood as the product of a unique sequence of development in which, as we have already noted, chance factors played a significant role in bringing about change. While Boas did not deny that all cultures had begun as hunter-gatherer ones, he maintained that, if the development of cultures displayed any regularities, these must be so subtle and complex as to defy understanding (Herskovits, 1953). Boasian anthropologists went to great lengths to undermine evolutionary claims by demonstrating that the historical record was much more complicated than parallel evolutionists had expected and that ethnographic traits that evolutionary anthropologists had believed were characteristic of widely separated stages of human development were regularly found together in the same living cultures (Mintz, 1985: 57). The only way to account for cultural change was to discover the successive diffusionary episodes that had shaped the development of each culture. While Boas was more optimistic about the potential for cultural development of American Indians than the racist cultural evolutionists of the nineteenth century had been, Boasian anthropology viewed aboriginal peoples as essentially imitative rather than creative.

Although largely German in its origins and outlook, Boasian anthropology shared some crucial ideas with an indigenous Euro-American anti-evolutionary movement that valued traditional aboriginal cultures even more highly than the Boasian anthropologists did. The North American naturalist Ernest Seton (1936) argued that, if one considered the quality of their traditional spiritual life and their relations with nature, aboriginal Americans were superior to Euro-Americans. Such views were encouraged by the Sierra Club, founded in 1892, and the National Parks Movement. These attitudes represented a continuation of the Rousseauian idealization of the primitive and the rustic and the still earlier contrast which the late seventeenth-century French adventurer Louis Armand de Lom d'Arce, Baron de Lahontan had drawn, based on his personal experiences among the Hurons and other North American Indians, between the soulless artificiality of French upper-class life and the Indians' spontaneity and closeness to nature (Sioui, 1992: 61–81).

Another school of diffusionist anthropology developed in Austria beginning in the late nineteenth century. The ethnologists of the Vienna School, founded by Fritz Graebner and later led by Wilhelm Schmidt, speculated, largely on the basis of comparative ethnographic evidence, that early cultural development had mainly been confined to Central Asia, a region that remained archaeologically largely unknown. As this Garden of Eden slowly became less habitable, cultures at various levels of development had been carried to different parts of the world. The complex patterns of cultures currently observed on different continents resulted from how far cultures at different levels of development had spread and the varying ways in which they had blended with one another in specific locales. The Vienna School also promoted the concept of primitive monotheism by trying to demonstrate that all hunter-gatherer cultures shared a belief in one god (Harris, 1968: 382–92). This effort by Roman Catholic scholars to discredit Tylor's evolutionary approach to religion represented a brief and unsuccessful resurgence of degenerationism within ethnology. Graebner and Schmidt tried to reconcile the archaeological evidence in favour of technological evolution with the prevailing pessimism about human creativity by claiming that evolution had occurred only once and primarily in one region. Theirs was in fact unilinear evolution in the literal meaning of that term.

A diffusionist approach replaced an evolutionary one in British ethnology as a result of the work of W. H. R. Rivers. When he was unable to detect an evolutionary pattern in his detailed field studies of the distribution of cultural traits in Oceania, Rivers rejected an evolutionary approach in favour of a diffusionist position. Diffusionism was carried much further by the Australian anatomist Grafton Elliot Smith, who became interested in mummification when he taught anatomy at the University of Cairo before moving on to the University of London. Noting that embalming had been practised elsewhere in the world, Smith decided that it had been invented in Egypt, where it also attained its most highly developed form. From Egypt it had been carried to many other places, where imperfect versions were introduced into local cultures.

Smith proceeded to theorize that all cultural development beyond the hunter-gatherer level had occurred in Egypt. Before 4000 BC there was no agriculture, architecture, religion, or government anywhere in the world. Then the harvesting of wild barley and millet on the Nile floodplain had led to the development of agriculture. Smith argued that this happened in a unique environmental setting which made the transition from gathering wild grains to planting them very simple and that it was extremely unlikely that agriculture would ever have been invented anywhere else. In fact similar settings exist in many other parts of northern Africa and opportunities for plant domestication seem to have been common throughout the world (Vavilov, 1951). Smith assumed that the invention of agriculture stimulated the development of pottery, cloth, monumental architecture, other arts and crafts, and divine kingship. These innovations were carried to all parts of the world by Egyptian prospectors and merchants searching for raw materials that had the power to cure diseases and prolong human life. While these acted as an 'exotic leaven', encouraging the development of agriculture and civilization in other parts of the world, Smith believed that derivative civilizations tended to decline when cut off from direct contact with Egypt (Smith, 1923). Degenerationism played an even more important role in Smith's thinking than it did in that of the Vienna School.

Smith's hyperdiffusionist ideas were elaborated, using ethnographic data, by William J. Perry (1923, 1924), who also taught cultural anthropology at the University of London. Perry documented worldwide parallels in political organization and religious beliefs, which he attributed to diffusion from the Nile Valley. Lord Raglan (1939) also advocated hyperdiffusionism but believed Mesopotamia rather than Egypt to have been its source. These scholars agreed that human beings were naturally primitive and would always revert to savagery unless prevented from doing so by the ruling classes; that savages never invent anything; that the development of civilization, and by extrapolation the industrial revolution, were accidents that produced results contrary to human nature; and that religious beliefs were the prime factor encouraging the development and spread of civilization. Yet, in denying that progress was natural or that there was any plan to human history, they were only carrying to an extreme basic ideas that had been shared by a growing number of anthropologists since the 1880s.

Relations between ethnology and the study of evolution did not improve when in the 1930s younger British ethnologists reacted against the sterility of hyperdiffusionism by rallying to the structural-functionalist approach of Bronislaw Malinowski and E. R. Radcliffe-Brown. Like the French sociologist Emile Durkheim, the new social anthropology viewed societies as integrated systems whose various institutions were linked together and functionally depended on one another like the organs of a living creature. Aspects of culture were said to acquire their significance only in terms of their functional relations to social systems. Social anthropology stressed the detailed investigation of individual societies, and the discipline as a whole was conceptualized as the comparative study of social morphologies, which made it analogous to comparative anatomy. Malinowski, and to a still greater extent Radcliffe-Brown, rejected all evolutionary and historical interpretations on the grounds that they were based on speculative and inadequate evidence and maintained that the comparative examination of the structure and functioning of the societies that were currently available for intensive ethnographic study was sufficient to encompass the morphological variation among all societies. It was not until the 1940s that E. E. Evans-Pritchard (1949, 1962) began to argue that observing how social structures changed made it possible to understand more precisely and objectively how their components were articulated.

Culture-Historicism and Archaeology

It is significant that while Boas, the Vienna School, and the hyperdiffusionists used archaeological evidence to support their positions, none of them attempted to use archaeological data systematically. Some European archaeologists were influenced by Smith to the extent that they argued that megalithic tombs might be degenerate versions of Egyptian pyramids, the idea of which had been carried to Western Europe by Egyptians who were searching for life-giving minerals (Childe, 1939: 301–2).

Yet, by the 1920s, the archaeological record was already sufficiently well known that hyperdiffusionism as an explanation of world prehistory had little appeal to archaeologists. While Ratzel had argued that diffusion had to be ruled out before the possibility of parallel evolution might be considered, archaeologists had already noted that New World civilizations were based on indigenous plants and animals that were totally different from those that had been domesticated in the Old World and that the earliest examples of their monumental architecture bore no closer resemblance to that of Egypt than did later phases. Most American archaeologists interpreted the prehistoric cultural development of the New World as having occurred separately from that of the Old World (Kidder, 1940; Spinden, 1928). Yet, under the influence of diffusionism, they did not discuss the significance of the separate development of agriculture, metallurgy, and civilization that was implicit in this position.

Already by the late nineteenth century, almost all change in the archaeological record, especially that which took place within the confines of major regions such as the eastern hemisphere, the Americas, and the circumpolar region, was being attributed in an anti-evolutionary fashion either to the diffusion of ideas from one group to another or to migrations that resulted in the replacement of one people and their culture by another. Western scholars studying China were convinced that a knowledge of agriculture, metallurgy, the chariot, and probably kingship had been introduced to the Huang-ho Valley from the Middle East (Fairservis, 1959). Those archaeologists who relied more heavily on diffusion tended to be not only more optimistic about human creativity but also more liberal politically than those who emphasized migration. Oscar Montelius, John L. Myres, Arthur Evans, and V. Gordon Childe all attributed the development of prehistoric European material culture largely to the diffusion of more advanced technology from the Middle East, although they also argued that ultimately Europeans had been able to use this technology more effectively than the peoples of the Middle East had. On the other hand, W. M. F. Petrie attributed all cultural changes in prehistoric Egypt to mass migrations and the arrival of smaller groups who effected sociocultural transformations by mingling culturally and biologically with the existing population. Petrie (1939) saw no possibility of significant cultural change without biological change.

It was also argued that hierarchical societies developed when politically dynamic pastoral peoples conquered less innovative peasant societies. Pastoral nomads were believed to be particularly adept at imposing their languages, beliefs, and cultural institutions on conquered peoples, while adopting the latter's material culture. Indo-European conquerors coming from the steppes of the Ukraine were imagined to have introduced a dynamism that vastly enhanced the creative potential of Europeans, while Hamitic conquerors were believed to have forced the rudiments of civilization, which they in turn had acquired from the Middle East, on the 'inert' Bantu peoples of sub-Saharan Africa (Evans, 1896; Mac-Gaffey, 1966; Myres, 1911). This was a transposition to prehistory of the belief that in the Dark Ages horse-riding warriors, largely of German origin, had imposed their authority over local peasants to become the ruling class of the medieval period.

It became fashionable for migrationists to assert that only an invasion from an external source could introduce a superior culture to a region. William J. Sollas (1911) saw little evidence of indigenous evolution but much evidence of migrating races in the archaeological record beginning as early as the Palaeolithic period. He viewed the superior might of a more gifted group as ensuring its military success and this in turn promoting the spread of biologically superior peoples and their cultures at the expense of less talented ones. Sollas further justified the colonial expansion of Europeans in North America and Australia by extending his principle that 'justice belongs to the strong' to mean that 'it is not priority of occupation, but the power to utilise, which establishes a claim to the land' (1911: 383). He regarded it as inevitable that superior races should drive weaker ones into poorer environments and that more demanding environments would nurture people who would dominate or exterminate the rest. The main danger threatening conquerors, especially if they were numerically small in relation to the people they dominated, was that through interbreeding the qualities that ensured their biological superiority might become diluted.

Formulations of this sort exhibited considerable ingenuity and subtlety. English historians and anthropologists were aware that Britain had been conquered and settled in turn by Celts, Romans, Saxons, Danes, and Normans and assumed that similar invasions must have occurred in prehistoric times. These scholars argued, however, that what was biologically and culturally best in successive invading groups had combined with what was best in the indigenous population to produce a people whose 'hybrid vigour', compounded of various European stocks, made them the best in the world (Rouse, 1972: 71–2). It was frequently maintained, for example, that the Norman talent for government in combination with Saxon love of freedom accounted for the development of parliamentary democracy in England (Freeman, 1867–79; Kemble, 1849; Stubbs, 1891–1903; Turner, [1799–1805] 1823). This formulation asserted the collective superiority of the entire British people, while at the same time biologically grounding a hierarchy in which the upper class was identified with the Normans, the English middle class with the Saxons, and peoples of Celtic descent with the still earlier and more primitive Britons. It also distinguished between the more centrally located Normans and Saxons and earlier peoples who 'survived' on the 'Celtic fringe'.

The extreme in a migrationary culture-historical approach to interpreting the archaeological record was the work of the German prehistorian Gustaf Kossinna (1911). Although trained in philology, he turned to archaeology in an effort to ascertain the original homeland of the Indo-European-speaking peoples and hence of his German ancestors. Inspired by fanatical patriotism, Kossinna declared archaeology to be the most nationalistic of sciences and the ancient Germans the most noble subject for historical research. Like Sollas and many other archaeologists, Kossinna believed that cultures were an expression of ethnic identity, which in turn was grounded in racial differences. He believed that, by identifying historically known tribal groups with specific archaeological cultures dating from the early historical period, it would be possible to trace these peoples back into prehistoric times. Kossinna concluded that all the Indo-Europeans could be traced back to the Mesolithic Maglemosian culture of northern Germany and Denmark. He also sought to demonstrate that this region had been the world centre of cultural development from earliest times. Even the alphabet, he proposed, had been invented there during the Stone Age rather than later in the Middle East.

Kossinna also accepted the distinction that the German ethnologist Gustav Klemm (1843–52) had drawn between culturally creative peoples (*Kulturvölker*) and culturally passive ones (*Naturvölker*). The only true *Kulturvölker* were the blond-haired, long-headed Nordics, or Aryans, who were the ancestors of all the Indo-European-speaking peoples. Because a more advanced culture was an expression of biological superiority, such a culture could only be carried from one region to another by a migration of people. Waves of Indo-Europeans were imagined to have spread south and east conquering native populations and using them to build civilizations in the Middle East, Greece, and Italy. Each of these waves had interbred with local populations, thereby impairing its creative capacities. Even the Indo-European-speaking peoples of Greece and Italy became less capable of sustained cultural creativity as a result of such behaviour. Because only the Germans had remained in their original homeland, they alone remained racially pure and therefore the most talented and creative of the Indo-European peoples. Only they were still capable of carrying out the historical responsibility of creating civilization and imposing it upon inferior peoples. This gave them the right to consider themselves the 'first-born' of the Indo-Europeans. Kossinna ignored or distorted archaeological evidence that did not support his position.

Kossinna's glorification of German prehistory as that of a biologically pure master race was intended to reinforce German nationalism and won the favour of conservative German politicians such as Paul von Hindenburg and later of some leading National Socialists such as Heinrich Himmler. Although Kossinna died in 1931, his view of German prehistory became a major component in the history curriculum that the Nazis imposed on German schools (Frick, 1934). By encouraging the Germans to regard the Slavs and other neighbouring peoples as inferior to themselves, Kossinna's theories helped to encourage aggression against these peoples.

In order to set Kossinna's ideas into perspective, however, two observations are essential. First, in portraying the other peoples of Europe as inferior to the Germans, Kossinna was not in principle acting differently from amateur and professional archaeologists in North America, Africa, and Australia who were portraying indigenous peoples as inferior to Europeans. Although the explanations differed somewhat, history and archaeology in each region reflected racist attitudes that had become widespread in Western civilization since the middle of the nineteenth century.

Second, it appears that Hitler did not subscribe to Kossinna's view of prehistory but regarded the ancient Germans as savages by comparison with their Greek and Roman neighbours (Speer, 1971: 141). Hitler's views were shaped not so much by German romanticism as by a form of social Darwinism that had been popularized by the German biologist Ernst Haeckel (Ingold, 1986: 52–3; Stein, 1988). This form of social Darwinism saw struggle as going on not so much among individuals or classes as among ethnic and racial groups. Haeckel believed that, as contending groups struggled to survive, tribe supplanted tribe and civilized nations supplanted less developed ones. Unlike more old-fashioned racists, who viewed biological superiority and inferiority as innate and unchanging (except as a result of racial mixing), Haeckel's followers viewed these attributes as responses to natural selection. For Hitler the source of German superiority lay not in the primordial past but in recent developments that had endowed them with the ability to dominate other peoples. Hitler died believing the Second World War was a test the German people had failed miserably. In his last days he sought, by destroying food supplies and other necessities, to outdo even the destruction resulting from total military defeat in order to ensure that the Germans were wiped from the face of the earth (Speer, 1971: 557, 560).

Soviet ideologues in the early 1930s denounced Kossinna's vision of history as the primary manifestation of the racist and reactionary spirit that animated the intellectual life of Central and Western Europe and North America. Their own extreme version of parallel evolution was presented as the one real alternative to such a view and as offering the only hope for humanity. Ironically, as the Soviet government came to rely increasingly on nationalism, and especially Russian nationalism, to rally opposition to German expansion in the late 1930s, Soviet archaeologists found themselves creating what came to be a Slavic version of Kossinna's prehistory.

Conclusion

Beginning in the 1880s, a concern with tracing the general, worldwide patterns of technological, institutional, and intellectual development gave way to an interest in studying the history of specific races, peoples, and cultures. British archaeologists and ethnologists succumbed to German romanticism and began to talk about cultures rather than treating culture as a unitary phenomenon as they had done previously. Boyd Dawkins identified Mortillet's Palaeolithic *époques* as the cultures of individual peoples and by the 1880s regional variations in the archaeological record were routinely being interpreted as the remains of prehistoric cultures and peoples. The concept of ethnic history corresponded with the intensifying nationalism of the period. As a result of growing racism, even less of an effort appears to have been made to distinguish between the biological and cultural factors that influenced human behaviour than had been made in the past. Sociocultural evolution once again became associated with revolutionary and left-wing politics, this time specifically with Marxism. More conservative individuals had become disillusioned with change and found highly congenial the ideas that people were naturally resistant to change and that much human behaviour was innate in the biology of human groups. Here too, however, there was a range of opinion, with more conservative scholars tending to believe that most human behaviour was innate and attributing change to migration rather than diffusion. On the positive side, the historical particularism of this period encouraged social scientists to pay more attention to studying variations in human behaviour and for the first time to examine individual cultures as entities worthy of detailed study in their own right.

7

Western Evolutionary Counterpoint

Historical particularism dominated the study of cultural change until the early 1960s. It could not, however, prevent archaeologists from perceiving general patterns in human history. Evidence of technological progress and of increasing sociocultural complexity continued to accumulate as archaeological work was carried out over ever larger portions of the globe. Yet racism, pessimism about the value of progress, and lack of faith in human creativity blunted an interest in the evolutionary implications of such findings. In general, archaeologists interpreted the past as a history of ethnic groups rather than of cultural change. Some Boasian anthropologists, such as Robert Lowie (1920, 1959), continued to deny that there was any overall pattern to human history and sought to demonstrate that the combinations of traits found in individual cultures might have been brought together by diffusion alone. So important were diffusion and migration as explanations of cultural change that all other views had to be justified in relation to them. This is nowhere more evident than in the work of the leading archaeological theorist of the early twentieth century, V. Gordon Childe, who struggled, not always successfully, to escape from the confines of the culture-historical approach.

Gordon Childe

Beginning in the 1930s there was a limited revival of interest in sociocultural evolution among Western archaeologists, anthropologists, and economists. This occurred as the economic crisis of that decade encouraged a growing sympathy for socialist economic policies and left-wing ideas generally. While some intellectuals explicitly embraced Marxist doctrines, others sought to ground policies of state intervention in a modified liberal philosophy, as embodied in the reformist ideas of the British economist John Maynard Keynes. The more radical Marxist option is evident in the later work of Childe. From the mid-1930s, he professed to be a Marxist archaeologist and his work was inspired first by Soviet archaeology and then by his investigation of Marxist philosophy. Childe published a number of studies of cultural evolution, two of which, *Man Makes Himself* (1936) and *What Happened in History* (1942), were widely read. Like Soviet archaeologists, Childe viewed the development of private property and class struggle as key factors bringing about sociocultural change. Yet his evolutionary works continued to reflect Western European reservations about human creativity and parallel evolution. He attributed differences in the development of civilization in the Middle East and Europe to ecological and demographic factors and to prehistoric Europeans being able to apply Middle Eastern innovations in a new setting. Within the Middle East he explained the differences between the city states that had developed in Mesopotamia and the divine monarchy that had united Egypt at an early stage in its cultural development in terms of alternative social and political techniques for controlling agricultural surpluses that had been devised in the course of the transformation of tribal societies into class ones. He appears to have regarded the different paths taken by these two civilizations as largely accidental (Childe, 1928, 1930).

For the rest of his life, Childe continued to argue that diffusion had played a major role in human history. He maintained that Scotland, which lacked domesticable plants and animals, would never have experienced a Neolithic revolution had already-domesticated plants and animals not been introduced from the south (Childe, 1946). In *Social Evolution* (1951) he argued that, even if different cultures that had the same mode of production tended, in accordance with Marxist theory, to evolve similar social, political, and cultural characteristics that reflected the influence of this base, such adjustments were constantly being disrupted by technological innovations that diffused from one society to another much faster than features of society adapted to the economic base. Childe concluded that, because of this, social reality rarely accords with an ideal type and sociocultural evolution tends to be multilinear and divergent rather than parallel. In his last book, *The Prehistory of European Society* (1958), Childe identified social relations, including those of production and control, as the main aspects of human behaviour that displayed cross-cultural uniformity and hence might be explained in terms of universal generalizations. He saw cultural traits acquiring their functional significance in relation to social systems and as being influenced by historical factors and by the vagaries of diffusion to such a degree that they had to be explained idiosyncratically. Thus, for Childe, embracing evolutionism did not involve repudiating a culture-historical approach.

Childe also showed more interest in why cultures did not change than why they did. He argued that at any stage of development, entrenched political hierarchies and inflexible religious beliefs could slow or even halt social and economic change. He distinguished between progressive societies, whose relations of production favoured the expansion of productive forces and in which there were positively reinforcing relations among the means of production, social institutions, and dominant systems of belief, and conservative societies, where the social, political, and ideological forces that Marxists designated as superstructure were able to block change. Childe believed that modern hunter-gatherer societies were ones that over many millennia had elaborated rituals and patterns of social organization rather than technology, and hence bore little resemblance to the progressive Palaeolithic hunter-gatherer societies that had evolved into modern civilizations

(Childe, 1942). Childe thus ascribed an important historical role to both the economic base and the sociocultural superstructure. Yet, in accordance with classical Marxism, he was careful to specify that where the superstructure was dominant, its influence could only be negative. The superstructure could block major sociocultural change, but it could not bring it about.

Childe argued more specifically that in complex societies, ruling classes sought as far as they could to prevent technological changes that might undermine their political authority. They did this by monopolizing most of the surplus wealth, exercising bureaucratic control over craftsmen, inhibiting the pursuit of greater technological knowledge, and patronizing the elaboration of magic, astrology, and other forms of superstition on a vast scale, as well as by the use of force. They only succeeded, however, in hanging on to power at the cost of making it more difficult in the long run for their own societies to compete with more progressive neighbouring ones. This explains why European societies were eventually able to outstrip and dominate those of the Middle East, despite the major technological advances that, during an earlier and freer period, had taken place in that region (Childe, 1942). Childe's view of Middle Eastern society had much in common with Marx's concept of the Asiatic mode of production, which Marx had described as a primitive form of class society that lacked sufficient internal dynamism to evolve further. This concept had been suppressed in the Soviet Union in the 1930s on the ground that it discouraged revolutionary activity in China and India (Bailey and Llobera, 1981; Dunn, 1982).

Childe also argued, much more explicitly than nineteenth-century evolutionists had done, that the history of individual civilizations did not follow a straight, upward path but took the form of bursts of energy and accomplishment followed by troughs of decline. When he wrote *What Happened in History*, Childe feared that the Nazis were dragging European civilization into another dark age. Yet he derived hope for the future by observing that in the past each new peak had risen higher than its predecessor, while no trough had ever sunk as deep as the one before. Still later, however, Childe observed that the archaeological record showed that progress was neither automatic nor inevitable and that many societies had suffered stagnation, decay, and extinction. Confronted by the spectre of the atomic bomb, he believed there was no certainty that Western civilization would progress in a rational manner, rather than 'vanishing' like that of the Classic Maya, or 'fossilizing' like that of the Chinese (Childe, 1947).

Although this combination of cyclical and linear concepts was not new, Childe was one of the first evolutionists to pay serious attention to the non-linearity of evolutionary progress and take account of sociocultural factors impeding change. His work exemplifies how far a Western Marxist committed to studying cultural evolution was influenced by the overall pessimism about human creativity that had pervaded European culture since the 1880s.

Evolutionary Revival

In the 1930s the rise of Nazism compelled many conventional Western anthropologists to question extreme forms of cultural relativism, which denied that it was possible to evaluate cultures against external criteria. Political realities made the notion that fascism could be judged only on its own terms seem immoral and unacceptable (Hatch, 1983: 103–4). The rise of Nazism also led anthropologists to assume a leading role in opposing racism. The discussions that were to culminate in the UNESCO Statement on Race published in 1952 (Shapiro, 1952) helped to resolve the confusion between race and culture that had perennially complicated discussions of human behaviour. They also helped to set cultural relativism into a broader perspective by reviving an interest in those characteristics that all human beings had in common as a result of belonging to the same species (Murdock, 1945). This, combined with major theoretical advances in biological evolution resulting from the progress that geneticists had made in understanding the nature of biological inheritance, encouraged various anthropologists and biologists to compare biological and cultural evolution in a more systematic fashion in order to determine in what ways these two processes were similar and different.

In 1939 the ethnologist Alexander Lesser argued that both forms of evolution were characterized by 'descent with modification' (Mintz, 1985: 81–3). While simpler cultures or species did not necessarily give rise to more complex forms, all cultures and species had long, interconnected histories, and advanced forms had inevitably developed from simpler ones. Lesser contrasted the many features that distinguished the 'simplest societies' from modern 'European civilization'. In the natural course of events composite tools had to be invented later than simple ones, metallurgy later than flint-knapping, and caste and slavery later than age and gender distinctions (1985: 85–9). In both biological and sociocultural evolution preceding conditions were necessary to explain consequences but not sufficient to produce them. Because of the influence of contingencies, both biological and sociocultural evolution tended to be multilinear. Lesser noted, however, that with cultural evolution the diffusion of ideas from one person to another was the most important mechanism bringing about change, while biological evolution was shaped primarily by natural selection. This made cultural evolution much more flexible, since individuals could change their minds about specific matters and create new combinations of ideas, while natural selection could operate only on the fixed combinations of genes that were present in individual organisms (Mintz, 1985: 92–9). Despite these valuable insights, Lesser was urged that, because the term evolution had such negative connotations, in his writings he should replace it with 'development' (1985: 92).

Similar ideas were elaborated by the anthropologist Alfred Kroeber (1952: 57–62) and the biologist Julian Huxley (1960: 56–84). Biological evolution was seen to occur as a result of natural selection acting on random genetic mutations. Over time species became reproductively isolated from each other, creating a dendritic pattern of evolution. Species that were well adapted to stable situations might remain constant over long periods, while others were rapidly overwhelmed by better-equipped competitors or transformed as natural selection adapted them to changing environments. Cultural evolution, by analogy, was seen to come about as the result of new ideas competing with older ones. But new ideas, unlike genetic mutations, were not created by a purely random process. Instead they were the products of rational, albeit culturally conditioned, human innovation. Kroeber (1948: 341–3, 364–7) moved away from Ratzel's and the young Boas's view that cultural innovation was a random process by noting that many scientists had made similar discoveries almost simultaneously and independently of one another as a result of working in the same cultural tradition. He thus inadvertently supported the Marxist contention that innovation is not an autonomous, but a socially conditioned, process. George Peter Murdock (1956) later argued that cultural innovations also resulted from inadvertent miscopying, which produced cultural drift, as well as from more deliberate forms of invention.

There was also a heightened awareness that cultural evolution was different from social evolution, which in part resulted from the growing importance of social anthropology. Culture came to be perceived as the concepts that guide human behaviour, society as the network of relations by means of which individuals interact with one another. It was suggested that every individual draws on available information to create a cultural map that guides his or her behaviour and that this map can be modified on the basis of personal experience or exposure to new ideas (Childe, 1956; Wallace, 1961). By contrast, social evolution takes place as technological and demographic change, or interaction with neighbouring societies, brings about transformations in economic, social, and political organization.

With respect to cultural evolution, it was accepted that far more ideas are learned from others than are invented by any one individual and that an infant becomes a functioning member of a society mainly as a result of what he or she learns in the course of socialization. Huxley and Kroeber observed that novel concepts spread within a society or from one society to another as the result of conscious choices. Individuals decide whether new ways of doing things or thinking about things are more or less effective and appealing than current ones. As a result of numerous individuals making the same choice, either for cultural reasons or out of practical self-interest, convergence plays a major role in cultural evolution. Neighbouring cultures may become more as well as less similar and some may totally integrate. The tree of cultural evolution, unlike that of biological evolution, is not dendritic; its branches can grow together as well as separate (Kroeber, 1952: 86, 93). Torsten Hägerstrand's (1967) more recent geographical analyses of the microprocesses involved in the spread of innovations in modern societies are a major contribution to the understanding of cultural change.

An important advance in investigating the reciprocal relations between social and cultural evolution was made by Murdock in his book *Social Structure* (1949), which sought to account for worldwide regularities in kinship terminologies. Using an extensive cross-cultural database, Murdock constructed a complex sequence of causation in which changes in social relations produced changes in culture. Specifically he argued that alterations in the sexual division of labour led to changes in postnuptial residence patterns, these in turn to changes in descent groups, and finally changes in descent groups to different classifications of kinship relations. Murdock also demonstrated that only six main systems of kinship terms possessed sufficient logical coherence to survive as stable types. His documenting of the numerous options that existed in organizing family life made it clear that there were more ways in which changes in social relations might come about than there were stable cultural outcomes in the lexical encodings of kinship relations. Later, in a comparative study of Neolithic societies in Europe and the Middle East, Gordon Childe (1951) demonstrated that, because of functional constraints, the degree of variation exhibited in the social and political organization of these societies was far less than that found in their general cultural patterns. Both Murdock and Childe concluded that the outcomes of change were more limited in variety than were the processes that brought change about. Childe demonstrated how this happened for functional reasons in the sociopolitical sphere and Murdock how it happened as the result of a need to maintain logical coherence in cultural patterns. It also appeared, however, that, in general, there was considerably less variation in social, political, and economic patterns than in cultural ones.

Another major contribution to understanding sociocultural evolution was Hornell Hart's (1959) 'Social theory and social change'. Summarizing earlier research, Hart identified three laws of cultural change. The law of cultural acceleration indicated that over the long sweep of time the power of human beings to carry out their purposes in the material, biological, psychological, and social realms tended to increase at an accelerating rate as knowledge and the social basis for knowledge grew more complex. The law of logistic surges postulated that progress with respect to specific goals is made slowly at first, then speeds up, and finally slows to a halt as the creative potential of specific innovations becomes exhausted. The law of cultural lag postulated that innovations lead first to changes in everyday behaviour and then to changes in institutions and belief systems in order to accommodate innovations and control their effects in the interest of society. Hart believed that at least the first two of these laws had been firmly established.

Clyde Kluckhohn (1960) presented a study listing twelve ways in which the moral order changed as societies grew larger and more complex. Among these he noted that morality became more explicit and abstract, more internalized, and better able to cope with ethnic and behavioural diversity. While he did not attempt to define stages of development, Kluckhohn delineated directional changes in morality that he believed accompanied the development of more complex societies.

As a result of this research, it became clear that there was considerable directionality in sociocultural change and that the processes that brought about change in the sociocultural and biological spheres were quite different. The ability of human beings not only to innovate but also to evaluate other people's ideas and either adopt or reject them meant that there was no clear-cut differentiation between the cultural equivalents of mutation and selection. This made cultural change much less of a blind and random process than was natural selection. While human knowledge is limited and unconscious habits may severely constrain innovation, sociocultural evolution is both more goal-driven and more efficient than is biological evolution. With its capacity to inherit acquired phenotypic adaptations, it resembles the purposeful Lamarckian evolution once envisaged by biologists (Ingold, 1986: 367–9). The work of this period thus resulted in the realization that Lamarckian mechanisms applied only to sociocultural and not to biological evolution.

It also became evident that, because sociocultural change is inevitably associated with changes in ideas, it occurs largely, although not entirely, independently of what happens to human beings as biological entities. A few ideas may have such disastrous consequences that their frequency is lowered as a result of the deaths of large numbers of people who chose to act upon them. For the most part, dysfunctional behavioural patterns are simply abandoned once their undesirable features are noted. Assuming that all humans, being members of the same species, have much in common, the fact that sociocultural evolution selects mainly ideas rather than individuals helps to explain why directionality and convergence are far more common in sociocultural than in biological evolution.

The more general and philosophically oriented evolutionism of this period continued to manifest the teleological and mystical qualities that had characterized sociocultural evolution in the nineteenth century. Curiously, however, the chief exponents of these views, which sought to present a reassuring account of humanity's place in the universe, were professional biologists rather than social scientists. The French palaeontologist Pierre Teilhard de Chardin ([1938–40] 1959) treated evolution as a process by means of which the cosmos gradually acquired intelligence in the form of the noosphere, which he defined as the interconnected totality of human minds and the knowledge they possessed. This concept reflected the most dangerous collectivist predilections of the political philosophies of both the extreme left and right of the mid-twentieth century. Julian Huxley (1960) saw human beings, through their acquisition of technological and scientific knowledge, becoming the chief agents of the entire evolutionary process, both sociocultural and biological, which under their control would cease to be shaped by blind selection and come to be guided by rational choice. He proclaimed that humans had become the 'only organism capable of further major transformation or evolutionary advance' and that their emergence had imposed severe limitations on the evolutionary possibilities of the rest of life (Huxley, 1960: 45). The destiny of human beings thus was to be 'the instrument of further evolution on this planet' (1960: 53). Huxley believed that for humans to play this role evolutionary thinking had to replace current religious and ethical beliefs and become the dominant approach to the integration of all ideas concerning human destiny and the shaping of general attitudes toward life. Huxley's views carried to an extreme the Enlightenment's faith in the capacity of human beings to manage themselves and the world in a rational manner.

While nineteenth-century evolutionists had attempted to provide a substitute for God by embedding purpose and meaning in the cosmic order, Teilhard de Chardin and Huxley treated human beings as the means by which the universe was gradually acquiring purposeful direction. This approach had much in common with Hegelian idealism. While their ideas elicited some public interest, they had little influence on the social sciences, which either remained loyal to cultural particularism or were trending towards a positivism that had little patience with mysticism.

Most Western studies of sociocultural evolution between the 1930s and 1960s had more modest goals. Yet they had important things to say about the mechanisms of cultural change. Attempts were made to measure the respective roles played by evolutionary and diffusionist processes in shaping trait distributions. In a study of data collected for all reasonably well-documented traditional North American Indian cultures, Harold Driver (1966) demonstrated that historical-diffusionist factors explained more about patterns of kin-avoidance behaviour than did existing functional-evolutionary explanations. He did not interpret this rather esoteric finding as negating the value of an evolutionary approach, but as helping to define the limits within which such an approach might be useful. This sort of sociocultural evolution no more rejected diffusion as a process bringing about change than nineteenth-century sociocultural evolution had done. On the contrary, having to contend with a prevailing culture-historical paradigm, it was even more respectful of diffusionist claims.

A number of recent sociocultural evolutionary studies are more closely aligned with the evolutionary concerns of the historical particularist tradition than with more recent neo-evolutionary approaches. One of these is Christopher Hallpike's *The Principles of Social Evolution* (1986). Hallpike observes that, while in small-scale societies many alternative arrangements seem to work equally well, there is a tendency for more complex societies to be more tightly integrated and generally more like each other than are simpler ones. Hence, even if they are internally more differentiated, early states are structurally more alike than chiefdoms or tribes and industrial societies are more alike than early states. Where competition is weak, smaller and less well-organized societies can survive forever (1986: 75–6). Yet, as competition grows keener and more societies come into contact with each other, less well-organized societies have less chance of surviving in the face of competition from more advanced ones.

Hallpike also stresses, however, that 'core principles' or beliefs, which form enduring elements of cultural traditions, play as important a role in shaping modes of production as does technology. While not denying the importance of material factors, Hallpike maintains that it is possible to predict more about the nature of a culture from knowing what language family it belongs to than from knowing the kind of natural environment in which it is found (1986: 370). This aspect of his work represents the elaboration of culture-historical themes that Kroeber (1952) had raised much earlier. While Hallpike's combined emphasis on functionalism and cultural traditions has much in common with classical Marxism, he resembles Kroeber in refusing to assign causal priorities to material factors.

Another major study in the historical-particularist tradition is *Culture and the Evolutionary Process* by the ecologists Robert Boyd and Peter Richerson (1985). Their primary concern is how culturally transmitted features interact with environmental contingencies to create the forces of human evolution. They maintain that selection is a key to understanding change in both cultural and genetic systems, but agree with previous analysts that cultural systems, unlike genetic ones, transmit acquired phenotypic adaptations, which take the form of beliefs and behavioural patterns. Boyd and Richerson argue that culturally transmitted traditions offer the most successful adaptive models for human behaviour. This explains why in, Francis Bacon's words, 'Custom is the principal magistrate of man's life' (1985: 81) and why humans for the most part inherit rather than choose their beliefs, as Sumner also maintained. Culturally transmitted ideas not only have been tested and refined over time by repeated applications to solving problems but are ones that, as a result of their success in maintaining themselves in a social context, have proved their ability to serve the interest of groups as well as of individuals. Traditional ways of dealing with problems work best in stable conditions. Yet, even when changing conditions necessitate innovation, human invention provides a far less reliable guide for behaviour. This, they argue, is because human reason is far more fallible than time-tested knowledge and because ideas tend to be formulated in terms of immediate personal interests rather than the long-term best interests of society. For the same reasons, slow incremental change is likely to be more efficient and less disruptive than is radical change. What Boyd and Richerson do not stress sufficiently is that rapid improvements in living conditions may make the cost of radical change seem worthwhile to economically and socially disfavoured groups. The eventual general benefits of new technologies may also outweigh the heavy social costs of their implementation.

Conclusion

Beginning in the 1880s the social sciences in Western and Central Europe repudiated evolutionism, embraced cultural relativism, and came to view cultural change as a largely random process. Human beings were thought to be uninventive and generally indisposed to change their way of life. Culture-historical anthropologists regarded cultures as random collections of traits that had come together as a result of idiosyncratic patterns of diffusion and migration. An alternative view was provided by Emile Durkheim, the British social anthropologists, and Talcott Parsons, who viewed social systems as functionally integrated entities. By the 1930s, even Boasian anthropologists had begun to view cultures, not as random collections of traits, but as patterns or configurations that were shaped by a human need for psychological and cognitive consistency.

All of these schools shared with nineteenth-century evolution an idealist view of culture as learned behaviour which changes as a result of ideas changing. Yet, while Parsons (1966, 1971) embraced an idealist concept of sociocultural evolution in which change was brought about mainly by 'normative elements', such as symbolic codes, legal norms, and religious and philosophical ideas, in keeping with current views that human beings were not inclined to change their ways, neither social nor Boasian cultural anthropologists exhibited much interest in studying change. Most social anthropologists followed Durkheim in viewing stasis as evidence of successful adaptation and social well-being. Despite this, by the 1930s, Boasian anthropologists such as Lesser and Kroeber were exploring the evolutionary facets of cultural change and, by making selective use of the concepts of biological evolutionists, were making significant contributions to understanding the nature and mechanisms of sociocultural evolution. Much of the writing about sociocultural evolution at this time was highly innovative, albeit rather dull and technical. Yet authors often shunned the label evolution, preferring to call their work studies of cultural change or development. The term evolution experienced a revival around 1959, in connection with the centenary of the publication of Darwin's *On the Origin of Species*. Even then, however, studies of sociocultural evolution sought to insert themselves into an ongoing culture-historical programme. While American archaeologists increasingly appreciated the extent to which aboriginal cultures had grown more complex in prehistoric times, most of them continued into the 1950s to attribute change mainly to diffusion and migration. Alfred Kidder had been a rare exception when he maintained that the aboriginal cultures of the prehistoric southwestern United States owed little more than their 'germ' to the outside and had developed locally and almost wholly independently of cultures in other regions (Kidder, [1924] 1962: 344). It is therefore not surprising that the study of sociocultural evolution within the culture-historical tradition was largely lost sight of as the investigation of evolution within American anthropology came in the 1960s to be monopolized by a materialist and explicitly anti-Boasian approach.

8
American Neo-evolutionism

The two decades that followed the Second World War constituted an era of unrivalled prosperity and political hegemony for the United States. They also witnessed renewed prosperity in Western Europe following the devastation of the war. All levels of society in the United States and Western Europe benefited from this prosperity, reducing class conflict (though not industrial disputes) to a minimum. Hence, despite the threat of nuclear war, this was a period of great optimism and self-confidence for the middle class. This optimism in turn encouraged a readiness to believe that there was an upward-moving pattern to human history and that technological progress was the key to human betterment. Within American anthropology these trends produced a new, more materialist approach to cultural evolution, which came to be called neo-evolutionism. While this approach never became the majority position in American anthropology, it grew increasingly popular during the 1960s, exerted significant influence throughout the discipline, and for a time dominated prehistoric archaeology.

Neo-evolutionism initially represented yet another effort by middle-class intellectuals to rationalize their increasingly privileged position by demonstrating it to be the inevitable and unsolicited result of evolutionary processes that allowed human beings to acquire greater control over their environment and ever more freedom from natural constraints. Yet neo-evolutionism was the product of a social environment that was economically and culturally very different from that of the mid-nineteenth century. Racism had been discredited by Nazi atrocities and was banished from respectable intellectual circles. The capitalism, focused on individual entrepreneurial activities, that had been lauded by Spencer and Sumner had given way to a new corporate capitalism. While small businesses persisted, the economy was now controlled by large corporations. This new capitalism, which relied far more heavily on managerial and bureaucratic skills and on team work rather than on individual initiatives, was idealized as the principal factor bringing about economic growth.

It is not surprising that neo-evolutionism differed in certain crucial respects from the parallel evolutionism of the nineteenth century. Its ecological, demographic, and technological determinism excluded the Enlightenment idea that cultural change occurred as the result of gifted individuals using their natural intelligence and leisure time to devise ways to exploit nature more effectively and improve the quality of human life. Most neo-evolutionists continued to maintain, as diffusionists had done, that human beings sought to preserve familiar styles of life unless change was forced on them by factors beyond their control. This position, now rationalized in ecosystemic terms, embodied views about human behaviour being naturally conservative that were far removed from the individual creativity that Spencer and most nineteenth-century evolutionists had used to explain cultural change. Neo-evolutionists also abandoned the notion that individual inventiveness offered an explanation of such human creativity as might occur. The idea that rationality and free will provided a basis for understanding human behaviour was a traditional Christian concept that had not been challenged by the Enlightenment, but it now appeared a poor explanation of human behaviour and history. In the course of the twentieth century, it had become increasingly fashionable for social scientists to explain human behaviour in terms of constraints imposed by the environment, demography, society, culture, or psychological factors. Many of these constraints were believed to be all the more effective as determinants of human behaviour because human beings were generally not conscious of them.

Forerunners

The two principal exponents of what would become neo-evolutionism were the American ethnologists Leslie White (1949, 1959, 1975) and Julian Steward (1955). Of these, White was the more traditional and theoretically conservative. Yet it was largely because of his efforts that by the late 1960s cultural evolutionism had regained 'considerable acceptance' in American anthropology (Carneiro, 1970a: 834). White regarded himself as the intellectual heir of Lewis Henry Morgan and violently rejected the historical particularism, psychological reductionism, and belief in free will that he identified as the major tenets of Boasian anthropology. In their place he offered 'general evolution'. Like most nineteenth-century evolutionary approaches, general evolution sought to understand progress as a characteristic of culture in general, though not necessarily of each individual culture. White used this perspective to ignore the influence of natural environments on cultures and of one culture on another. Instead he concentrated on explaining what he regarded as the main line of cultural development. This line was marked by the most advanced culture of each successive period of human history, regardless of where it was found or what historical connections might or might not exist between it and earlier or later leading cultures. White's evolutionary sequence therefore consisted of a series of cultures, each of which was more complex than the one which had preceded it. In many cases the most advanced culture bore no direct historical relation to the previous most complex one. White's approach strongly privileged increasingly complex grades of sociocultural organization as the principal object of evolutionary study. He maintained that this approach was justified because cultures that failed to keep abreast were superseded and eventually absorbed by their more progressive neighbours. Hence from a long-term, evolutionary point of view they were irrelevant.

White conceptualized cultures as elaborate thermodynamic systems that adjusted populations to the natural environment. Each cultural system was composed of a techno-economic, social, and ideological subsystem. Throughout most of his career he argued in a utilitarian vein that cultures exist to make human life more secure and enduring and hence to benefit individuals. In 1975, noting the extent to which war and violence had characterized human history, White rejected this view as being unduly teleological and argued that cultures evolved to serve their own needs. His perception of how cultural change occurred was materialistic and narrowly deterministic. He maintained that 'social systems are ... determined by technological systems, and philosophies and the arts express experience as it is defined by technology and refracted by social systems' (White, 1949: 390–1). His basic law of evolution stated that, all things being equal, culture becomes more complex as the amount of energy harnessed *per capita* increases, or as the efficiency of putting energy to work is improved. He summarized this law with the formula

$$C = E \times T \text{ (Culture = Energy} \times \text{Technology).}$$

According to White, cultures that increase their energy throughputs will dominate, absorb, and destroy ones that are less successful at doing this. Yet, despite the sweeping claims that White frequently made for his ideas, he stressed that, although they accounted for the general outline of cultural development, they could not be used to infer the specific features of an individual culture (White, 1945: 346). Presumably he believed that this could be done only by means of detailed historical studies.

White's technological determinism is frequently alleged to be Marxist in origin (Sanderson, 1990: 90–1). White was an ardent socialist who published numerous political articles in left-wing periodicals under quixotic pseudonyms such as John Steel (Peace, 1988). Yet, apart from a shared materialist orientation, his ideas have little in common with classical Marxism. Instead, they reflect a traditional American tendency to privilege relations between technology and society at the expense of other relations, such as those between individuals and society (Kroker, 1984: 12). This view reflected the experiences of a relatively small Euro-American immigrant population that controlled a resource-rich continent. These experiences encouraged them to conclude that technological development provided the key to all other forms of success.

Steward championed a more specifically ecological and empirical approach to the study of sociocultural evolution. Like previous evolutionists he assumed that sociocultural development displayed significant uniformities, but he also believed that ecological adaptation played a crucial role in influencing the development of cultures. Steward therefore sought to determine how cultures had developed in different environmental settings. He postulated that cultures in similar environments, such as tropical forests, arid river valleys, or boreal forests, anywhere in the world would independently develop similar basic characteristics and follow similar developmental trajectories. On the other hand, cultures in different natural settings were likely to develop along different paths. Steward labelled this view of sociocultural evolution multilinear, and contrasted it with White's general evolution and the unilinear (parallel) evolution of the nineteenth century.

Steward maintained that similar features shared by cultures that were at approximately the same level of development and had evolved in similar environmental settings constituted empirical evidence of a 'cultural core'. The core embraced economic, political, and religious patterns that had major adaptive significance. Many of these features related to everyday activities, such as providing subsistence and shelter. Steward argued that the aim of evolutionary anthropology should be to explain the features shared by all or most cultures at the same level of development, rather than trying to account for the 'unique, exotic, and non-recurrent particulars' that were produced by historical accidents (Steward, 1955: 209). By this he meant that it was important to explain the development of major institutions such as kingship, which are found in numerous societies around the world, but it was of no scientific value to ascertain why kingship was symbolized in one place by a crown and in another by a stool, parasol, or ritual seclusion. Steward emphasized a tendency already present in American anthropology to identify sociocultural evolution with the study of cross-cultural regularities, and history with the study of particularities. He also encouraged a tendency to view evolution and history as mutually exclusive concepts and to regard only the study of evolutionary phenomena as capable of producing scientific generalizations. In this fashion, Steward sought to purge anthropology of Boasian historical-particularism. If anthropology was to become a science, it had to stop explaining specific instances of cultural change, as was done by historical particularists, and instead must devote itself to explaining cross-cultural regularities. Boasian anthropologists, such as Kroeber (1952: 63–78), had long used the terms evolution and history, or science and history, to distinguish between generalizing and particularizing explanations of human behaviour, but they had regarded these two sorts of explanations as being complementary rather than antithetical and of the same rather than of different degrees of importance.

Although Steward labelled his approach multilinear evolution and argued that cultures developed differently in different ecological settings, he continued to maintain that pristine civilizations could only arise in arid regions where irrigation agriculture had to be utilized to support expanding populations (Steward, 1949). Only in these locations, he argued, following Karl Wittfogel (1957), was it possible for the techniques required to manage irrigation to provide a foundation for the construction of the state. In all other regions, civilizations developed only as a result of the impact that early irrigation civilizations had on less-developed neighbouring societies. Wittfogel also believed, however, that in arid regions the limitations of water supplies imposed ultimate limits on the development of civilizations, which did not exist in many less arid environments. Steward's macroevolutionary scheme, which there is no evidence he ever abandoned (cf. Carneiro, 1973: 95–6; Sanderson, 1990: 93–4), was thus effectively a unilinear one which required the parallel development of early civilizations in arid regions of the world. Multilinearity appears to have been a concept that Steward applied to technologically less-developed societies and possibly also to secondary civilizations.

Early Neo-evolutionism

White and Steward were lonely voices in the 1940s and 1950s, but by the early 1960s sociocultural evolution had become a popular alternative to a culture-historical understanding of human history. Marshall Sahlins and Elman Service (1960) tried to reconcile White's and Steward's approaches by differentiating between general and specific evolution. General evolution was characterized as being concerned with increasing complexity, while specific evolution paid attention to ecological adaptation. It was argued that these were complementary concerns, each of which was necessary to produce a rounded understanding of cultural change.

Although this formulation dissociated sociocultural evolution from inevitably implying progress, in their later evolutionary studies Sahlins (1968) and Service (1962, 1975) used ethnographic data to construct a speculative and highly generalized sequence of unilinear development that characterized societies as evolving from bands of hunter-gatherers through tribes and chiefdoms to states. The specifics of the second and third stages were derived largely from Sahlins's (1958) and Service's (1962) detailed knowledge of the indigenous societies of the Pacific region, with Melanesian big-men societies playing a major role in defining tribes and the more hierarchical Polynesian societies providing a model of chiefdoms. It was assumed that under propitious conditions big-men societies of the Melanesian type would evolve into chiefdoms. Although it has been objected that an approach that systematically compared societies around the world would have revealed much more diversity than Sahlins and Service had assigned to each of these stages, their evolutionary sequence was widely accepted as an accurate portrayal of significant cultural diversity. Underlying this scheme was the assumption that cultural variation was quite limited.

Sahlins and Service both argued that increased productivity resulted in greater specialization, higher levels of social integration, and societies that were able to adapt more easily to their environments. Morton Fried (1967) posited a similar unilinear sequence running from egalitarian to ranked, stratified, and class societies. His scheme emphasized the role that political factors played alongside economic ones in bringing about the development of more complex societies. In each of these schemes the main emphasis was on how societies could progress from lower to higher stages of development. There was also some discussion about whether, and under what circumstances, it might be possible for a society to skip a stage. Sahlins and Service (1960: 93–122) believed that this was possible, but only if a more evolved society was already present to provide a model for change.

Although the primary emphasis in these neo-evolutionary schemes was on accounting for how societies evolved from one broadly defined stage to another, change itself was treated as occurring slowly and gradually. This view was clearly opposed to the Marxist argument that social change took the form of sudden, revolutionary resolutions of tensions that had built up over long periods of time between different classes in society. American archaeologists also objected to Childe's metaphorical use of the term revolution to designate such major transformations as the development of agriculture (the Neolithic revolution), the origin of states (the urban revolution), and the emergence of an industrial economy (the industrial revolution). In an era of McCarthyism and 'Red baiting', revolution was a sufficiently dangerous concept that academics had to distance themselves from it to avoid any taint of Marxism. They did this by attributing sociocultural evolution largely or wholly to ecological factors that operated from outside society and by ignoring or rejecting the Marxist idea that conflicts within societies played a significant role in bringing about social changes.

Theoretically the most sophisticated and durable version of neo-evolutionism was Marvin Harris's (1979) cultural materialism. Harris was less concerned with delineating evolutionary trends than were Sahlins, Service, and Fried, his principal aim being to explain how sociocultural change occurred. Nevertheless, he believed that parallel and convergent developments were more prominent features of sociocultural evolution than were divergent ones. Harris assigned a privileged role in shaping cultural systems

to an array of material conditions, including technological, demographic, ecological, and economic relations, and sought to explain
all sociocultural phenomena in terms of the relative costs and
benefits of alternative strategies as measured in terms of these
criteria. For Harris the most powerful constraints on human
behaviour were at the basic (infrastructural) level and it was
these constraints that shaped domestic and political economies
(structure), as well as values, ideas, symbols, and rituals
(superstructure). Structural and superstructural elements did,
however, play vital system-maintaining roles. Much of Harris's
work has been directed towards explaining the origins of food
taboos, religious beliefs, and cultural esoterica by discovering the
roles these customs play in relation to basic economic considerations (Harris, 1974, 1977). Unlike Steward or White, who left the
explanation of much of the cultural content of specific societies to
culture-historians, Harris has tried to demonstrate that these items
are shaped by economic considerations.

Another significant early contribution to neo-evolutionary
thought was Robert Carneiro's circumscription theory. Carneiro
(1970b) argued that early states could only develop in areas where
the land that farming people could cultivate was strongly delimited
by less productive areas of mountains, seas, or deserts. As population densities increased and led to conflict over possession of land,
people were compelled to submit to domination and exploitation
as a result of having nowhere else to go. Still other neo-
evolutionists stressed the importance of resource diversity and
long-distance trade as stimuli for the development of economic
specialization and eventually of political control (Renfrew and
Cherry, 1986; Sabloff and Lamberg-Karlovsky, 1975; Steward,
1960).

Kent Flannery (1972) equated cultural evolution with a society's expanding capacity to process, store, and analyse information
and argued that the types of sociopolitical organization that could
develop at any particular level were strictly limited by the arrangements that would permit those in charge to collect sufficient data,
make decisions, and ensure that those decisions were carried out.
Flannery suggested that, while a wide range of ecological, social,

and cultural factors might promote sociocultural evolution, its trajectory was determined mainly by the limited types of sociocultural systems that could function adequately at any particular level of complexity. Like Murdock (1949) and Childe (1951), he believed that the factors causing change and the specific ways in which change might come about were considerably more diverse and less determining of the course of sociocultural evolution than were the arrangements that are workable at any given level. This approach clearly viewed functional explanations as complementary to evolutionary ones.

Lewis Binford (1962, 1983) played a major role in adapting neo-evolutionary concepts for interpreting archaeological data. He defined the major goal of archaeology as being to formulate generalizations that would explain the cultural changes that are documented in the archaeological record. The most important questions archaeologists had to address were how some hunter-gatherer societies had evolved into food-producing ones, and some of these in turn into states. Binford, like White, viewed cultures as adaptive systems composed of technological, social, and ideological subsystems, which provided the energy required to sustain a human society. He described evolution as 'a process operative at the interface of a living system and its field' (1972: 106) and interpreted changes in all aspects of cultural subsystems as adaptive responses to alterations in the natural environment, population density, or adjacent and competing societies. This ecosystemic view ruled out human inventiveness as an independent force bringing about major changes. It also treated cultures as normally moving toward equilibrium or homeostasis, with changes being induced by external factors that tended to impact first on technology and then through technology on social organization and value systems. Yet Binford, like other neo-evolutionists, believed in the capacity of human beings to invent new forms of technology, social behaviour, beliefs, and values when adjustments were required by various external factors. Steward (1955: 182) had already argued that every cultural borrowing ought to be construed as if it were an 'independent recurrence of cause and effect' and Harris (1968: 377–8) had dismissed diffusion as a 'nonprinciple' when it came to explaining sociocultural change. Thus Binford and the other neo-evolutionary anthropologists emphasized humanity's capacity to innovate at the same time as they were maintaining that cultures which were not being challenged by environmental changes, demographic shifts, or external competition tended to be static. Pessimism about the value of creativity had eroded faith in human inventiveness in the 1880s, while not challenging an older belief that sociocultural change resulted from the efforts humans make to enhance their control over the environment and improve the quality of their lives. In neo-evolutionism a material causality was combined with a belief that people do not wish for sociocultural change but have the ability to innovate when forced to do so. For Binford the main forces that stimulated human creativity were located in the natural realm and shaped human behaviour even though human beings were often unaware of them.

While Binford and other neo-evolutionists proposed new mechanisms of cultural change, they made few deliberate attempts to test the validity of their theories. This was strange, because the majority of anthropologists did not subscribe to the materialist causality that neo-evolutionists were advocating. Rhetoric and selected examples were substituted for a systematic review of the issues, and evolutionists and anti-evolutionists seemed to rely on the future to vindicate their respective positions. Binford has inadvertently provided the most interesting test of the adequacy of early neo-evolutionary theories of change. He has spent much of his career trying to discover reliable ways to infer human behaviour from archaeological data (Binford, 1977, 1981). He argues that, because human behaviour represents rational responses to ecological challenges, such behaviour should display a high degree of cross-cultural uniformity, which would be especially evident among societies at the same stage of development. This led him to postulate that it ought to be possible, by studying living societies, to establish many strong and highly specific regularities between material culture on the one hand and various forms of economic behaviour, social organization, and beliefs on the other that would permit these forms of behaviour to be inferred from the archaeological record. Yet, despite the early hopes that he and his followers had for the success of this method, it has not produced the large numbers of correlations they expected, except with respect to those aspects of human behaviour, such as technological processes, where natural and biological constraints provide clear insights into how people did things in the past. Binford's middle-range generalizations tell us little more about religious beliefs than was known thirty years ago. This suggests that most spheres of human behaviour display far less uniformity than Binford and other neo-evolutionists believed.

Theoretical Diversification

Despite neo-evolutionism's enduring commitment to a materialist explanation of cultural change, neo-evolutionary understandings began to diversify considerably in the 1970s. Some neo-evolutionists have sought, as far as possible, to account for cultural change using the same concepts that biologists employ to explain biological evolution (Dunnell, 1980; Maschner, 1996; O'Brien, 1996; Teltser, 1995). They argue that, while human behaviour is learned and symbolically mediated to a far greater degree than that of any other species, culture is merely another mechanism for facilitating ecological adaptation; hence it performs the same functions as do the behavioural patterns of every other species and can be analysed using the same concepts. This sort of approach radically transforms the way in which social scientists pose questions. For example, David Rindos (1984) has defined domestication as a mutualistic relation between different species. He maintains that there ought to be no difference between explaining how plants and animals adapt to human needs and how human beings adapt to the needs of plants and animals. Cognitive processes are treated as strictly instrumental and epiphenomenal.

Robert Dunnell and his followers have criticized traditional cultural evolution for failing to internalize such key tenets of scientific (biological) evolutionary theory as random variation and natural selection (Dunnell, 1980; Wenke, 1981). While admitting that the mechanisms of trait transmission are more varied and the units on which selection operates are less stable with respect to cultural than to biological phenomena, they maintain that an approach based on the general principles of biological evolution will produce explanations of human behaviour that are more economical and insightful than those resulting from cultural evolutionary theories.

Linda Cordell and Fred Plog (1979) sought to apply a Darwinian approach by arguing that in every culture a wide range of different ideas is present in people's minds, including many ideas that are mutually contradictory. The ideas that tend to survive and spread are those that prove most adaptive in specific situations. Cordell and Plog also assumed that cultures that lacked the ideas necessary to survive died out, leaving few traces of their short existence in the archaeological record by comparison with more successful ones.

This approach, which seeks to treat human beings in the same way as it does other animals, carries to an extreme the neo-evolutionary denial that consciousness and intentionality play a significant role in shaping human behaviour. Most social scientists object that this approach is unduly reductionist and ignores the major differences already established between biological and cultural evolution. Already by the 1950s, it was recognized that the analogy between biological and sociocultural evolution broke down for various reasons. In biology the source of genetic variation is mutation, which appears to be a random process at least in relation to selection. Environmental factors favour the survival of genetic mutations that adapt the individual to its milieu, a process that clearly has no fixed direction. By contrast, cultural innovation often takes the form of a conscious modification of existing knowledge to achieve a desired result. Hence cultural innovation is to some extent purposeful, goal-oriented, and therefore, at least in the short term, teleological. Chance and social circumstances influence innovation, but innovation is related to the needs of individuals and groups far more directly and efficiently than is mutation. By drawing on personal experience and general knowledge to envisage the results of alternative strategies, human beings can often avoid dangerous and disruptive behaviour without having to resort to a costly process of trial and error. An approach modelled on biological evolution thus fails to take account of the efficiencies introduced into human relations with the environment and with each other by the human ability to symbolize.

Natural selection also occurs at the level of the individual organism. Hence a single, randomly mutating gene can significantly influence the chances of an individual organism surviving and reproducing. Because human beings learn from one another, sociocultural selection generally operates on individual ideas rather than on individual organisms. The capacity for human beings to change their minds and alter their behaviour has thus created a form of selection that operates far more quickly and efficiently with regard to individual organisms than does natural selection.

Boyd and Richerson (1985: 288–9) also argue that socially selected and transmitted knowledge generally provides better-tested models for individual human behaviour than do individual decisions. One of their objections to sociobiological interpretations of highly specific forms of human behaviour is that sociobiologists ignore the relative costliness and susceptibility to error of individual reasoning. Boyd and Richerson believe that, given the accelerating rate of cultural change over the past 10,000 years, understanding the details of cultural transmission is essential for understanding the evolution of human behaviour (1985: 14). Their findings strengthen the conclusion of evolutionary studies of the culture-historical period that the differences between the processes of sociocultural and biological evolution are far more profound than are the similarities (see chapter 7 above).

Other neo-evolutionists moved in a different direction by adopting general systems theory to account for sociocultural change. Efforts were made to determine how sociocultural systems actually worked by empirically quantifying the flow of goods and energy between one part of a social or cultural system and another and between such systems and the natural environment. The concept of energy was especially favoured as a unit of measurement as it was compatible with a broader ecological approach (Hosler et al.,

1977; Thomas, 1972; Watson et al., 1971). Moreover, the cost of moving goods, information, or anything else could be measured in terms of energy expenditures. Systems theory allowed analysts to transcend the limitations of traditional social anthropological analyses of static structures by studying not only structure-maintaining but also structure-elaborating (morphogenetic) processes.

Many of the most important general systems studies have utilized cybernetics, which seeks to account for how systems function in terms of positive and negative energy feedbacks among their component parts. Negative feedback maintains a system in an essentially steady state in the face of fluctuating external inputs by carefully regulating energy flows from one part of the system to another. The model of such behaviour is a thermostat, which keeps the temperature of a room at a nearly constant level by regulating heat flow. Another example would be the classic Malthusian relation between population and a fixed food supply. Positive feedback seeks to explain how uncontrolled shifts in energy flows can bring about irreversible transformations in the organization of systems. The creation of states may produce conditions that encourage the expansion of rural populations beyond the level where these populations can survive without the political security provided by state structures. That in turn becomes a factor that reinforces and ensures the survival of the state (Watson et al., 1971).

The concept of feedback offered social scientists a more precise and quantifiable means for establishing the relations among the different parts of a social or cultural system than did the essentially static concept of functional integration, and also made it easier to account for change. This approach encouraged a return to a more inductive and informative study of how sociocultural systems operated. Yet, as was the case with neo-evolutionism generally, most of those who adopted it continued to give priority to a practical knowledge of material processes and calculating the costs of alternative courses of action in relation to energy flows. Art, values, and religious concepts continued to be dismissed as being of only epiphenomenal status and having little importance in bringing about sociocultural change.

Other forms of systems theory focused on the managerial costs of social systems. These were measured in terms of the energy expended processing the information required to regulate such systems. It was observed that, as societies became more complex, specialized systems were required to process information and to make and implement the decisions that were required to manage them. This encouraged the development of hierarchical structures, control of which exempted key figures from the constraints that public opinion exercised over leaders in small-scale societies and permitted these key figures to enhance their power and wealth at the expense of their followers (Flannery, 1972; Johnson, 1973). As information-processing systems grew more elaborate, however, they tended to consume increasing amounts of wealth. This, it is argued, led to efforts being made to curb expenses, first by simplifying and standardizing decision-making and administrative procedures and then by decentralizing them and encouraging regions and groups to manage their internal affairs at their own expense (Rathje, 1975). On the basis of such thinking, it has been argued that the more complex economies of late pre-Hispanic Mexico were more entrepreneurial and free of state control than they had been in the earlier stages of the development of indigenous Mexican civilization (Blanton et al., 1981). Yet historical and archaeological evidence of very intimate links between economic and political activities in late prehistoric Mexico call into question whether overall management costs were being reduced at this period, or rising administration costs were being shared by a growing number of groups within Mexican society (Hassig, 1985; Parsons et al., 1982).

A third trend in neo-evolutionary explanations of sociocultural change, also of a systemic nature, was inspired by the work of the anthropologist Roy Rappaport (1968, 1979). While continuing to analyse cultures as adaptive systems, Rappaport stressed the role of ideology and ritual in helping to regulate economic activity and maintain feedback that promoted cultural stability. He treated beliefs as an inheritance from the past that provides societies with socially esteemed guidance and suggested that draping nature in supernatural veils offers it some measure of protection against human folly and extravagance. This accords with Boyd and Richerson's (1985: 274–7) argument that under stable conditions group selection tends to favour ideas that enhance the cultural success of the group over that of the individual. While even deeply held beliefs can be modified as conditions change, it is evident that some patterns of belief persist for long periods and play an important role in reinforcing social solidarity and cultural identity. This approach to human ecology is the only one that acknowledges cultural beliefs as playing a major role alongside rational calculations in shaping human behaviour. As applied in the work of Kent Flannery and Joyce Marcus (1976), this approach encourages an understanding of the past that combines the best of a neo-evolutionary approach with the judicious employment of a culture-historical perspective.

Neo-evolutionism was greatly strengthened as a result of the invention of radiocarbon and potassium-argon dating and the worldwide expansion of archaeological activity following the Second World War. For the first time it became possible to correlate prehistoric sequences everywhere and to consider the rates as well as the direction of cultural change. In most areas cultural change turned out to have occurred far more slowly than archaeologists had believed (Renfrew, 1973). This made it seem more likely that significant cultural development had resulted from local innovation rather than from diffusion. Archaeological findings also confirmed and expanded Nicholai Vavilov's (1951) contention that there had been many different centres of plant domestication, and the belief that early civilizations had developed slowly and largely independently of one another in many parts of the world. In Africa the Palaeolithic sequence turned out to be much longer than archaeologists had imagined and it became doubtful whether there had been any significant change in the earliest Oldowan stone-tool tradition over a period lasting a million years or longer.

Neo-evolutionary theory remained materialist and adaptationist in orientation, but over time its explanations of cultural change grew steadily more divergent. This resulted from the failure of neo-evolutionists to find evidence of the high degree of uniformity in human behaviour they had expected. As a result, many of them assigned a greater role in shaping human behaviour to historical traditions and to what materialists regard as epiphenomenal factors. Early neo-evolutionism reaffirmed the traditional evolutionist concern with progress, which it conceived as following an essentially unilinear path. The principal difference was that progress was no longer interpreted in idealist terms as resulting from human beings employing their powers of reason to control nature more effectively and improve the quality of their lives. Instead humans were viewed, much as they had been during the culture-historical period, as changing their lifestyle only when compelled to do so. Culture was interpreted from a materialist perspective as an adaptive system and change was thought to occur only when imbalances developed among the material factors that were involved in this relationship. Reason, which had been regarded as the motor of change, was now viewed merely as a sort of calculator that helped to bring about ecologically necessitated changes. Within a materialist framework the alternative explanation of cultural change was that of Marxist evolution, which attributed it mainly to interest groups within societies struggling to control limited resources. By downgrading the role of both social relations and individuals as agents of change, neo-evolution was forging a view of progress that was more in accord with the corporate capitalism of the 1950s than was the parallel evolutionism created in the context of the more individualistic early industrial capitalism of the nineteenth century.

Cataclysmic Evolution

In the 1970s the neo-evolutionary paradigm began to experience a major change. Since the late 1950s, the optimism and security of the American middle class had been seriously eroded by a succession of chronic and deepening economic crises that were exacerbated by failures of foreign policy, especially in Vietnam. Eventually, expanding economic difficulties began to erode public optimism among the Western European middle class also. This undermined faith in the general benefits that were to be derived from technological development. Once again, as had happened in Europe in the late nineteenth century, technological development, which had been viewed as the principal source of betterment for humanity, was accused of being the major source of human misery. In the short term, this did not undermine a belief in sociocultural evolution. Instead it produced a version of sociocultural evolution that was still further removed from Enlightenment ideals than early neo-evolutionary theory had been.

The new version of sociocultural evolution grew out of a series of middle-class protest movements that reflected a growing mistrust of technology and of the direction in which modern society was thought to be moving. These movements profoundly altered social values and had a major impact on the social sciences. Each is, or has been, politically powerful, but presents itself as dealing with a scientific issue of public concern rather than addressing major political and economic questions.

The oldest of these is the ecology movement. It is concerned about the ways in which unrestricted technological development is poisoning and destroying the world ecosystem. Its beginnings were signalled by the publication of Rachel Carson's *Silent Spring* (1962), a book that drew attention to the insidious dangers of environmental pollution. The ecology movement has since promoted concern about the immediate dangers to public health that are posed by a broad array of technological processes. The assault on the biosphere that has been documented includes ozone depletion, the greenhouse effect, acid rain, desertification, pollution of air, soil, and water by toxic chemicals, reduction in the number of plant and animal species and in biotic diversity generally, and the disruption of carbon-dioxide, soil-nutrient, and hydrological cycles (Nash, 1995). Environmentalists warn that in the long run even more catastrophic consequences may result from continuing pollution of the environment. Widespread concern about such processes has encouraged an increasingly wary attitude towards technological change and unrestrained economic growth (Ashworth, 1986; Bellini, 1986; Dotto, 1986; Fyfe, 1985; Jacquard, 1985; Leiss, 1990; Myers, 1984; Richardson, 1990).

The second movement, which sought to promote a conserver

society, stressed that key natural resources that were essential for industrial development, such as oil, gas, and fresh water, were available in nature only in finite quantities. Hence the world might rapidly be approaching a point where further industrial expansion would become impossible or the exhaustion of specific resources might result in declining living standards or even the collapse of industrial civilization. Conservation and recycling were advocated as ways to delay this impending crisis (Keating, 1986; Polgar, 1975). Hitherto, evolutionists had assumed that new raw materials and sources of energy would be discovered before old ones were used up. This faith in technological solutions to problems posed by resource depletion was so unquestioning that it had been believed there was no point in conserving resources that technological progress would render obsolete.

Paul Ehrlich's *The Population Bomb* (1968) drew public attention to a third major cause for anxiety. He argued that, if the unprecedented population growth made possible by modern infectious disease control were not checked quickly, the ecological and political costs would be disastrous. Rapid population growth in poorer countries would undermine their chances of economic development, leaving them and the planet increasingly vulnerable to violence and political chaos (Freeman, 1974; Laughlin and Brady, 1978; Polgar, 1975).

These movements raised issues that made social scientists and the public in North America and Western Europe increasingly sceptical about the benefits to be derived from technological progress. As political and economic insecurity increased, they came to view technological development as a source of danger and perhaps ultimately of disaster, as the Western European middle class had done in the late nineteenth century. Rapid change of any sort was condemned as a source of dysfunctional 'future shock' (Toffler, 1970).

These developments encouraged a shift in attitudes towards sociocultural evolution. The resulting paradigm marked a further retreat from the optimistic view of change created during the Enlightenment, and intensified neo-evolutionism's rejection of the belief that technological innovation was the result of a process of rational self-improvement and the principal factor that promoted cultural development. Two specific findings, in economics and social anthropology, encouraged this shift.

The Danish economist Ester Boserup (1965, 1981) argued that, while increasingly intensive modes of agriculture yield more food per unit of land and hence support denser human populations, they require increasing amounts of labour for each unit of food that is produced. Hence, according to Boserup, only the need to support slowly but inevitably increasing population densities would have led human groups to adopt more intensive forms of agriculture. The idea that population pressure might be a driving force promoting cultural change was not a new one. Herbert Spencer had believed that it had encouraged the development of agriculture and many other improvements in production, as well as the spread of human groups (Sanderson, 1990: 26–7). As early as 1673 the British statesman William Temple had argued that increasing population densities forced people to work harder (Slotkin, 1965: 100–11), and in 1843 the Swedish archaeologist Sven Nilsson (1868: lxvii) had maintained that increasing population had resulted in a shift from pastoralism to agriculture in prehistoric Scandinavia. Early in the twentieth century Raphael Pumpelly (1908, 1: 65–6), Harold Peake and Herbert Fleure (1927), and Gordon Childe (1928) had argued that food production had originated when postglacial desiccation in the Middle East had compelled people to cluster around surviving sources of water, where they had to develop more intensive forms of food production in order to feed higher population densities.

Yet these older generations of scholars had viewed the development of food production, however it came about, as an advance that permanently benefited humanity by providing food surpluses that made daily life more secure, supported denser clusters of people, and permitted greater craft specialization and more leisure time. Boserup's findings suggested that food production originated as a response to factors beyond human control. Throughout history the inability of humans to prevent population increase had resulted in them having to work harder, suffering from increasing acquisitiveness and exploitation, and degrading the natural environment to an ever greater extent. Marvin Harris generalized Boserup's argument by maintaining that the primary reason people created more advanced technologies was not to make their lives easier and more enjoyable but to stop them from getting worse. As populations increased, the only effective response was to work harder and adopt technological innovations that were increasingly expensive to maintain (Harris, 1977). This was the antithesis of earlier evolutionary arguments that technological and scientific advancement resulted in ever more effective control and exploitation of the environment for the benefit of humanity. This change corresponded with Leslie White's (1975) disillusionment with the idea that culture existed to satisfy human needs.

The demonstration by the anthropologists Richard Lee and Irven DeVore (1968) that hunter–gatherer societies could support a low population density with less effort than was required by even the least demanding form of food production not only supported Boserup's position and extended it to cover the whole of human evolution (Cohen, 1977), but also inspired a major reinterpretation of prehistoric hunter–gatherers. Instead of being regarded as living on the brink of starvation, they were portrayed as leisured groups that had plenty of time to devote to religious and intellectual pursuits. The traditional view of hunter–gatherers as living brutish and short lives was replaced with an image of 'primitive affluence' (Sahlins, 1972). Anthropologists also began to portray more egalitarian cultures as examples of 'conserver societies', that provided models of how Western peoples should think about and treat the natural environment.

Some anthropologists questioned the ethnographic evidence on which these views were based, especially Boserup's claims that more intensive forms of food production were invariably more energy-expensive in relation to each unit of food that was produced, and that population increase was an independent variable (Bronson, 1972; Cowgill, 1975; Harris, 1979: 87–8). Sharply dropping birth rates in Western countries, as a result of more effective methods of birth control, new work patterns, and the rising cost of caring for children, suggested that on the contrary birth rate was a dependent variable, at least in modern societies.

Anthropological evidence of widespread birth-spacing practices, abortion, and infanticide also suggested that population increase had been subject to human control throughout history. Yet the rapid and, for a long time, relatively unchallenged manner in which these studies influenced the interpretation of archaeological data, often in the absence of adequate ways to measure population size or even relative changes in population, indicates the extent to which these ideas accorded with the spirit of the time.

Archaeologists began to express their own reservations about conventional neo-evolutionary theories which treated sociocultural change as if it occurred only in slow, gradual trajectories of the sort that Robert Braidwood (1974) and Richard MacNeish (1978) had documented in the 1950s in their studies of the origins of food production in the Middle East and Mesoamerica. On the basis of his examination of long-term changes in settlement patterns in southern Iraq, Robert Adams (1974) pointed out that some very abrupt changes had occurred in the organization of early civilizations, which were separated by long periods when there were few changes or changes happened very gradually. Parallels began to be drawn between discontinuous cultural change and punctuated equilibrium in biology.

American archaeologists also became increasingly sensitive to evidence of decline in the archaeological record. Some examples were attributed to natural disasters, such as volcanic eruptions and unusual floods, or to human activities such as warfare or over-exploitation of the environment (Yoffee and Cowgill, 1988). The decline or disappearance of populations was recognized as occurring more commonly than had hitherto been believed. In previous decades gaps and discontinuities had been smoothed over by stretching finds backwards and forwards in time or by postulating yet-undiscovered cultures. The collapse of the Old Kingdom in Egypt, the disappearance of the Classic Maya civilization in the lowlands of southern Mexico and Central America, and the disintegration of the Roman Empire had been noted in the past, but had not been identified as processes requiring an evolutionary explanation. Instead the emphasis had been on overall patterns of development and the passing of the torch of civilization from one people to another. This in part reflected evolutionary anthropology's antipathy toward romanticism and romanticism's preoccupation with specific cultures. Steward (1949) alone had explained cycles of prosperity and collapse in early civilizations within an overall evolutionary framework. He interpreted downturns as resulting from populations expanding beyond their ecological limits and then collapsing as a result of ensuing conflicts over resources.

Decline and collapse are now recognized as aspects of change in the archaeological record that need explaining no less than do technological and other forms of advance. This realization encouraged Colin Renfrew (1978) to apply catastrophe theory, the mathematical study of how sets of variables within a system can produce discontinuous effects, to account for discontinuities in the archaeological record. Joseph Tainter (1988) has argued that in class societies the demands of bureaucracies and privileged elites habitually expand to the extent that they undermine the economy that sustains them, leading ultimately to the collapse of the upper levels of control or even of the entire society. Such studies of collapse reinforced the belief that societies generally are more fragile and sociocultural change more fraught with danger than most evolutionists had acknowledged.

A similar pessimism pervaded other evolutionary studies of this period. The sociobiologist Edward O. Wilson (1978) argued that only during the last few thousand years had the rate of cultural evolution exceeded that of genetic evolution. Noting that the biological traits that had ensured success for hunter–gatherers might not be suitable for coping with life in more complex societies, Wilson interpreted the speeding-up of sociocultural evolution as a potentially dangerous development. This conclusion is, of course, based on his conviction that successful human adaptations are biologically grounded in a highly specific rather than a general fashion.

A more interesting socioculturally based example of pessimism is Richard N. Adams's elegant *The Eighth Day: Social Evolution as the Self-Organization of Energy* (1988). Like his teacher, Leslie White, Adams argues that energy flows provide a basis for measuring and explaining sociocultural change. Societies are conceptualized as dissipative structures naturally tending towards entropy, and within which every energy conversion has a cost. Because of this, social change can be analysed in terms of the natural selection of self-organizing energy forms and of triggering processes, many of them symbolic, that regulate the flow of energy within each social system. Cultures are maintained by expending energy to ensure that the cognitive models of individuals conform at least minimally with a general pattern that ensures a reasonably efficient use of energy. Larger and more complex social systems tend to dominate smaller ones, even though the *per capita* costs of maintaining larger ones are significantly higher. Hence large systems have significant competitive advantages over smaller ones as long as free energy can be harnessed. The greater regulatory needs of larger systems also ensure the growing hierarchization of power and decision-making. In recent centuries expanding reliance on non-human energy sources has resulted in growing exploitation of the environment. It has also resulted in human concerns becoming relatively less significant in social planning.

Adams does not view social evolution as a mechanical process imposed on human beings, or as something that is predetermined and inevitable simply because it is subject to substantial natural constraint. The ultimate problem, as he sees it, is not so much where in the future new sources of energy might be found but where the tapping of new sources of energy might lead humanity. In his view, an endlessly expandable source of energy would result in 'ever greater complexity and indeterminacy, producing non-linearities beyond the coping ability of the human intelligence' (1988: 241). Thus, while he does not view the outcome of human history as wholly determined, Adams is pessimistic about the long-term consequences of technological evolution.

Pessimism about the future is not new. It has been deeply embedded in speculative thought since almost as far back as the earliest written records and seems to be a common trait of complex societies. But, until the 1970s, evolutionism and pessimism appeared to be antithetical modes of thinking about human history. In the 1970s new ideas about the nature of sociocultural change promoted a pessimistic and even tragic version of cultural evolution that saw demographic, ecological, and economic factors constraining change to move along lines that human beings do not regard as desirable but are unable to alter. This new, cataclysmic evolution implied that the future was likely always to be worse than the present and that humanity was journeying from a primitive Eden, filled with happy, leisured hunter–gatherers, to a hell of mass poverty, oppression, ecological disaster, and perhaps thermonuclear annihilation. Neo-evolution differed from nineteenth-century sociocultural evolution in its rejection of the belief that cultural change occurred as a result of human beings employing their powers of reason to control their natural environment more effectively. Cataclysmic evolution saw human beings locked into a process in which technological elaboration could only increase human misery. Sociocultural evolution had come to have much in common with its former antithesis, degenerationism.

Staying on Top

Bowler (1986: 236) has suggested that a loss of faith in progress was encouraged as a consequence of the definitive rout of anti-Darwinian ideas of biological change by the modern synthetic theory of biological evolution. That involved the final rejection of the Lamarckian idea that biological evolution might acquire a direction as a result of purposefully acquired biological characteristics being inherited from one generation to the next. Such a suggestion does not, however, explain the specific directions that were followed in turn by neo-evolutionism and cataclysmic evolution. These differed markedly from previous disillusionment with progress, which had resulted in the denial that there was any fixed order to human history. Instead, neo-evolutionists continued to posit a fixed trajectory for sociocultural development that human beings might hope to slow or halt, but not alter, and that in the case of cataclysmic evolution would sooner or later lead to general ruin. Few continued to argue that it was possible to act effectively to redirect the course of human history.

This view of evolution reflected one of the self-interested, and hence unacknowledged and largely unconscious, assumptions of the various movements that were trying to curb unlimited technological development and population increase. All these groups sought to avoid disaster by halting or severely slowing down the rate of change. Their idea seems to have been that each society and class would continue to occupy the same position that it presently did in the world economy. Thus the third world would pay the cost of halting progress, while those who lived in technologically advanced countries would retain their relatively privileged positions. For members of the middle class, whose position was being threatened by shifts in the world economy, halting change seemed to be a way to protect their past achievements.

One of the positive intellectual accomplishments of neo-evolutionism was that progress, and especially technological progress, was no longer treated as a force that automatically brought about social and moral improvement. On the other hand, by ascribing the direction of sociocultural evolution to external forces and denying a role to human volition and decision-making, neo-evolutionism had created a new mystique that was at least as powerful and potentially as dangerous as the mythology of automatic moral progress had been.

It is frequently claimed that neo-evolutionists, like the parallel evolutionists of the nineteenth century, regarded defining the shape of human history as being more important than explaining why it happened. This is probably true, inasmuch as the explanations of change that were offered in both cases tended to be simplistic and reductionist. Yet each phase of evolutionary theorizing produced its own distinctive explanation of how sociocultural evolution came about. The ecological and demographic determinism of neo-evolutionism reflected declining faith in the ability of the individual to influence a mass society, while cataclysmic evolution embodied growing doubts that technological change would continue to favour the interests of those who had benefited most from the existing order.

The main tactical weakness of neo-evolutionism was its uncompromising and confrontational attitude towards other schools of thought. Its exponents failed to take account of the ideas of culture-historical anthropologists, even when these ideas were evolutionary ones. Many such ideas complemented those of neo-evolutionism and, if they had been taken seriously, could have helped to produce a more rounded and powerful synthesis. The refusal to consider them created gaps in neo-evolutionary theory that prepared the way for yet another theoretical backlash. It may only be, however, through confrontation and antagonistic debate that all the various implications of different positions can be worked out and appreciated.

9
Evolution Attacked Again

In recent decades technological change has been occurring at an ever-accelerating rate. In developed countries it has become increasingly difficult for most people to understand on what principles the technology that has become essential for their daily life is based. A few decades ago individuals were able to repair their own cars and mechanical typewriters. Today only specialists can replace the modular components in automobiles and few users of electronic word processors understand even in general terms how these machines operate. At the same time, the economy has become increasingly internationalized, investment more specialized, and the management of personal finances more complicated and perplexing. When combined with major economic changes that dislocate the lives of individuals, increase unemployment and personal debt, and shatter most people's hopes of long-term stability in their lives, rapid technological and economic change produces dismay as well as a growing sense of helplessness. All this makes a mockery of the enhanced personal freedom that Durkheim (1893) predicted would result from the increasing interdependence brought about by a growing division of labour.

Once again, as happened in earlier times when technological changes created social and political difficulties, artists and scholars have turned to idealism and romanticism for solace, and the concept of evolution finds itself in a hostile intellectual environment (Giddens, 1981, 1984; Mandelbaum, 1971; Nisbet, 1969).

In the late nineteenth century it became fashionable to romanticize earlier and apparently less confusing periods of human history and to emphasize stability in human behaviour by viewing it as biologically determined and resistant to change, as well as highly variable from one human group to another. Various ethnic, national, and community groups sought to enhance identity and meaning in their lives by emphasizing their biological and ethnic roots. For them ethnic history became more important than studying the general progress of humanity. Today, as individuals grow increasingly dependent on economic systems that they feel ever less able to control, writers and philosophers are glorifying the essential independence of the human spirit from the material circumstances in which individuals find themselves. They stress the power of the human mind to shape the perception of reality and, through that perception, individual behaviour. In this tendentious fashion, these intellectuals seek to deny the material and social constraints that press upon their lives and attempt to reassert the freedom of the individual without having to confront material and political realities in any direct fashion. This negating of material constraints by denying that they have power is the essence of postmodernism.

Postmodern Critiques

Postmodernism, the late twentieth-century incarnation of romanticism, has spread through the social sciences from comparative literature and literary criticism. It stresses the subjective nature of knowledge, embraces extreme relativism and idealism, and promotes critical theory as a means to unmask the self-interest that is alleged to parade as objectivity within all received understandings. Extreme forms of postmodernism maintain that there is no objective knowledge: every decoding of a message is another encoding and what claims to be truth is in reality a means to defend and promote personal and collective power and vested interests. All established approaches, with the possible exception of postmodernism itself, are represented as serving specific interests (Hunt,

1989; Laudan, 1990; Rose, 1991; Rosenau, 1992). The credibility of such ideas has been boosted in anthropology by Ian Hodder's (1982) demonstration that material culture, apart from its technological significance, is not a passive reflection of social organization but plays an active role in the negotiation of social relations, often distorting and disguising as well as reflecting social reality.

Postmodernism also celebrates random idiosyncratic cultural variation and, as an idealist practice, tends to reject proposals that human behaviour is shaped to any significant degree by external physical constraints (Geertz, 1973; Goffman, 1963; Knorr-Cetina, 1981; Latour and Woolgar, 1979; Shepherd, 1993). Postmodernism has reinvented the culture-historicism that prevailed in the early twentieth century. While it has much to say that is of value and was lost sight of during the positivist period that followed the Second World War, its extremism threatens to recreate rather than transcend the problems of the culture-historical approach. In particular, postmodernism makes it impossible to gain insight into the origin, structure, and change of social systems at a time when massive changes in these systems are affecting everyone's life, often for the worse (Sherratt, 1993: 125). There is a real danger that those who ignore the history of the social sciences will be doomed to repeat the errors of the past.

Extreme postmodernists and critical theorists argue that sociocultural evolution is a myth that has been created to make the privileged position of the middle class seem to accord with the natural order of things and rationalize the colonialism, repression, exploitation, and abuse of nature from which the middle class has benefited (Diamond, 1974; Giddens, 1984; Rowlands, 1989). It is argued that not all societies are prone to change and that those that are evolve along many divergent paths (Golson and Gardner, 1990). Historical surveys (such as this one) are said to demonstrate that sociocultural evolution is a self-serving concept that lacks objectivity and restricts the questions scholars ask and the range of explanations that appear reasonable. This renders the concept bankrupt scientifically as well as politically and morally. The goal of progressive scholars, it is maintained, should be to deconstruct sociocultural evolution's claims and expose its mythic status. What is not stated is that, since extreme postmodernism implies that no concept has other than mythic status, postmodernists are singling out evolution for criticism entirely on political grounds.

It is also argued that progress is an illusion. Arno Karlen (1995), following William McNeill (1976), has sought to demonstrate that sociocultural progress since the Neolithic period is nothing compared to the progress that has been made by human diseases. His book documents the vast amount of human suffering brought about by human population growth, more intimate contact with animals and animal products, and widespread malnutrition. The spread of new diseases is currently being facilitated by the intrusion of resource-hungry economies into hitherto untouched natural environments and by the ever-increasing and ever-faster transportation of people and goods around the world (McMichael, 1995). Ecological disasters caused by natural forces, abuse of the environment, and political catastrophes are also seen as creating more human misery than has been acknowledged in the past. Evolutionary optimism has been challenged by many who would agree with Eric Hobsbawm (1994: 13) that the twentieth century has been 'without doubt the most murderous century of which we have record'. It is fashionable to believe that simpler is better and that in Palaeolithic times human beings had already reached the limits of cultural development that were appropriate to their biological and psychological adaptation. These readings of human history represent a continuation and deepening of the disillusionment with progress that created cataclysmic evolution and helped to redress the naive optimism of earlier times. Yet the basis of that earlier optimism was not a denial that much was wrong with human existence at all times, including the present, but the belief that, in spite of this, human numbers kept increasing (which was then seen as a good thing), technology became more complex, and total knowledge continued to expand. All of this, it was hoped, would result in a better future.

Postmodernists argue that it is ethnocentric to claim that one society (usually the analyst's own) is superior to, or more evolved than, any other; a view which can only excuse and promote racism, colonialism, and imperialism, as well as deny the value of cultural diversity and the extent of human creativity. All modern societies are products of equally long periods of human experience and there are no objective grounds on which it can be established that one society's views or behaviour are superior to those of another. Others go further and maintain not only that each group has the right to interpret its past, or human history generally, in its own way (which is perfectly acceptable) but also that there is no method for evaluating the resulting histories in relation to one another (Brose, 1993: 13–14). This claim denies that divergent interpretations can be evaluated against factual evidence.

The concept of progress likewise is being denounced as ethnocentric and ideologically motivated. Anthony Giddens and others have maintained that it is empirically incorrect to interpret most of human history as a 'world growth story' (Giddens, 1984; Shennan, 1993). It is also argued that it is wrong to describe some groups as being more backward than others. The relativist Boasian claim has been revived that, while traditional Australian aborigine societies may have been less advanced technologically than were European ones, their kinship organization was more complex (Tilley, 1995). Giddens (1984: 236) has argued that 'Human history does not have an evolutionary "shape", and positive harm can be done by attempting to compress it into one.' He observes that there is a mistaken tendency to interpret particular paths of historical change as if they were general and universal and to equate the superior military, economic, and political power of Western societies with moral superiority (1984: 242). It has also been charged that evolutionary research often concentrates on evidence that supports its theories, while ignoring evidence that does not.

During the twentieth century only Soviet evolutionists claimed that all societies developed in a precisely similar manner. Indeed a growing concern with societies as adaptive mechanisms militated in favour of evolutionists paying increasing attention to specific differences. Processual evolutionists now acknowledge that cultures develop along different lines in different regions and among different language groups (Flannery and Marcus, 1983). There is also in relativist critiques much confusion of ethical considerations with factual ones. It is rarely mentioned that extreme relativism precludes not only the measuring of all societies against any common standard but also the possibility of ascertaining whether any general improvements have occurred over time. This approach in effect seeks on general principles to preclude any findings contrary to the ones that its proponents favour.

Some of the most vitriolic objections have been directed against claims that perceptions of reality have changed in some systematic fashion as technologies and societies have grown more complex. It is implied that the ideas on this subject of Lucien Lévy-Bruhl (1922), Henri Frankfort (1948; Frankfort et al., 1949), Gordon Childe (1949, 1950, 1956), Christopher Hallpike (1979), and Ernest Gellner (1988), if not racist, err by treating adults in non-Western societies as if they thought and behaved as children do in Western civilization (Giddens, 1984: 239–40). In fact very little of the writing on this subject claims that, when it comes to particular matters involving an understanding of cause and effect, the analytical abilities of one group are significantly different from those of another. Nor does it have to be maintained that the ways people think vary in any essential way between small-scale and large-scale societies. Even if the linguistic concepts that encode logic differ from one culture to another, the structure of categorial reasoning appears to remain the same (Hamill, 1990). What does appear to be systematically different is how reality is perceived and the rigour with which it is analysed as intellectual life becomes more specialized. Early societies saw the natural world as impregnated with supernatural forces that possessed wills and intelligence similar to those of human beings. One of the great developments of human history has been the gradual realization that the natural world has to be understood and related to in a totally different way from the social realm if both are to be dealt with objectively. This was the great discovery of what Karl Jaspers called the axial era of Eurasia in the mid-first millennium BC (Eisenstadt, 1986). It has been argued that even something as axiomatic as mathematics or formal logic has been compelled to change since the time of Aristotle in order to accommodate itself to new types of empirical data (Darwin, 1938). To ignore developments of this sort on the ground that not to do so is hegemonous and anti-relativist is to overlook one of the major developments that has transformed human understanding and behaviour.

Gordon Childe (1949: 6–8) stressed that no people adapts to the environment as it really is but to the environment as their culture leads them to imagine it to be. Yet he added that a society's understanding of what the environment is like has to correspond fairly closely to reality if their adaptation is to be successful and that society is to survive. Cholera can be controlled more effectively by burning garbage than by burning witches (Childe, 1956: 58–60). Childe thus provided a basis for judging the relative effectiveness of cultural systems of knowledge as adaptive patterns.

Ethnologists who studied the subsistence patterns of the Cree hunter-gatherers of northern Quebec in connection with the James Bay hydroelectric development project of the 1970s documented the detailed knowledge that these people had acquired of their environment, which greatly exceeded that possessed by Euro-Canadian scientists. Much of this knowledge was encoded in a belief system that conceptualized relations between hunters and game animals in terms of spiritual relations between humans and animal spirits. These beliefs had sustained a highly successful relation between the Crees and their environment for several millennia. Yet the Crees' superior knowledge did not suffice to answer the question that was most important to them in the 1970s: how much land would they need to retain control of in order to ensure that all Crees who wished to go on being hunter-gatherers might do so? Answering that question required setting Cree knowledge into the analytical framework of modern ecology, which had no place for Cree beliefs in animal spirits (Feit, 1978; Salisbury, 1986; Tanner, 1979). This demonstrates that what distinguishes knowledge at different levels of social complexity is not necessarily the quality of knowledge but the frameworks within which it can be manipulated. Such frameworks transform knowledge into an increasingly powerful instrument not simply for maintaining but for consciously transforming societies and their relation to the environment.

While the intellectual achievements of small-scale societies must not be belittled, neither must the technological significance of changing human knowledge as societies have grown more complex. Yet these achievements are less impressive when considered from a moral perspective. As a result of general evolution, individuals do not appear to have become any less inclined to sacrifice the lives and happiness of others to protect and expand their own wealth and power. The twentieth century has witnessed the mass slaughter of individuals and groups for ideological and ethnic reasons that ultimately appear grounded in the perceived material interests of their oppressors. Yet the sacrifice of human beings can no longer be justified on the grounds that such sacrifice is necessary to sustain the cosmos or to make crops grow, and burning witches is no longer seen as an effective way to control epidemics. Thus a more realistic understanding of the natural order has eliminated the rationales for many cruel and irrational practices and resulted in many parts of the world in a better quality of life as measured by increasing numbers of human beings, increasing longevity, and improved levels of health and education. All of this represents the realization of some of the basic goals of the Enlightenment.

Neo-historicist Critiques

It has also been objected that neo-evolutionary stages do not constitute a set of specific social formations but are vaguely defined categories whose features overlap in a disconcerting fashion. Many real societies display combinations of traits associated with both Sahlins's big-men (or tribal) societies and with chiefdoms, and Melanesian chiefs, so defined by their kin-ranked and inherited positions, often have as their foremost goal to become big men by achieving super-trader status. There is also no evidence of there having been a transition from big-men societies to chiefdoms as societies grew larger in Polynesia; on the contrary, chiefs appear to have been present even within the first small bands of colonists in that part of the Pacific Ocean (Yoffee, 1993: 64–5).

It is also objected that unilinear developmental schemes ignore the extent of variation that is found among societies assigned to the same stage. No effort has been made to document systematically the geographical variations of this sort, although it is abundantly clear that African chiefdoms were different in many crucial respects from Polynesian ones and that the structure of North American Indian tribes was radically different from that of New Guinean peoples. One crucial difference is the almost universal refusal of New Guineans to create viable communities of more than a few hundred people (Forge, 1972). It is further argued that, by emphasizing the importance of centralized hierarchies, a unilinear approach has diverted attention from the full range of control structures in early states. Recent studies have sought to define a series of alternative arrangements in which shared power can integrate states as effectively as does centralized control (Blanton et al., 1996). It is not clear to what extent these sorts of variations can be explained either ecologically or culturally.

Finally, increasing doubt is being cast on the assumptions that the types of societies associated with neo-evolutionary stages constitute a unilinear evolutionary trajectory and that 'any representative of a given cultural stage is inherently as good as any other, whether the representative be contemporaneous and ethnographic or only archaeological' (Sahlins and Service, 1960: 33). Morton Fried (1975) suggested that tribal societies are products of colonial encounters, which caused hitherto independent villages to form larger and more complex societies in order to resist aggression. Since then it has been questioned whether bands always give rise in turn to tribes, chiefdoms, and states, or whether tribes, chiefdoms, and states can each arise directly out of band-type societies. Norman Yoffee (1993: 73) has also questioned how alike are a chiefdom society that is rapidly evolving into a state, such as that found at Teotihuacan, Mexico, during the Formative period (300 BC–AD 150), and a chiefdom society where ecological limitations preclude further evolutionary development, such as seems to have been the case in early historical Hawaii. Yoffee suggests that there was much less real inequality in Formative Teotihuacan than in Hawaii, even though the potential for inequality was much greater. Over fifty years ago and for quite different reasons, Gordon Childe (1936: 44–7) had called into question the evolutionary assumption that early societies were essentially similar to modern ones at the same level of development.

Today there is growing interest in the impact that intersocietal relations have on social change. It is argued that it is wrong to study individual societies in isolation from other ones with which they were in contact (Renfrew and Shennan, 1982). Formerly, studying societies in isolation had been justified by the neo-evolutionist assumption that a similar ecological adaptation produced a similar society and culture, as well as by Julian Steward's (1955: 182) argument that, regardless of whether new traits originated inside or outside a society, every change in a society's developmental trajectory could be interpreted as an 'independent recurrence of cause and effect'. Yet even Steward (1955) distinguished between the development of primary and secondary states, while Fried (1975), as we have already noted, treated many of the features of modern tribal societies as responses to colonial aggression. Today, primarily as a result of the work of Eric Wolf (1982), it is being argued that all modern pre-industrial societies must be viewed as having been transformed in significant ways by colonial encounters prior to the earliest ethnographic descriptions. Not all of these encounters were with Western societies. This strengthens the implication that less-developed societies in the modern world are not precise analogues of what societies were like prior to contact with more complex neighbours. Many anthropologists object that Wolf is too rationalist and Eurocentric in his approach and hence fails to appreciate the resilience of indigenous cultures in the face of such encounters (Vansina in Whitehead, 1995: 313–14). Those who hold this position, however, are generally romantics, who see cultural diversity as being inherently more idiosyncratic than the rationalistic Wolf does.

Efforts have been made to endow studies of societal interaction with greater theoretical rigour by applying world-system theory, which attempts to understand how in recent centuries peripheral or semi-peripheral areas, which tend to provide raw materials and markets, relate to the core areas of regional or worldwide economies, which specialize in industrial production and finance (Wallerstein, 1974). World-system theory purports to provide a set of concepts that yield insights into the processes by which societies at different levels of economic development influence one another. Even for early civilizations, however, it appears that the differences between cores and peripheries were not as clearly defined economically as they are in the modern world. Long-distance trade was mainly in luxury goods and was intended for elite consumption. The concept of world system may have to be altered considerably before it is appropriate for the study of premodern times (Rowlands et al., 1987).

It is also clear that hunter–gatherer societies have survived to the present only in marginal and ecologically very demanding areas. Elsewhere, most of them long ago either evolved into more complex societies or were replaced by groups that had agricultural economies. The complex collecting cultures of California and coastal British Columbia are among the few that survived long enough to be investigated in detail by anthropologists. Even when hunter–gatherer societies have survived into modern times, it is possible that they were significantly modified as a consequence of contacts with more developed societies prior to being studied. Some hunter–gatherer societies may have been significantly altered as a result of being able to rely on neighbouring agricultural ones for food. It is not known, for example, to what extent the ways of life of aboriginal societies in northern Canada were altered over the past several centuries as a result of being able to trade furs for imported food at Hudson's Bay Company posts. Some anthropologists argue that because their aboriginal way of life was highly flexible, their cultures were not significantly altered by such contacts (Francis and Morantz, 1983). Others postulate a shift from big-game hunting to trapping and the use of more base camps and specialized hunting parties, as well as the development of family hunting territories, as consequences of access to the resources obtained at year-round trading posts (Ray, 1974). It is also being questioned whether the San (Bushmen) of South Africa were pristine hunter–gatherers or the descendants of pastoral groups that had lost their cattle (Schrire, 1984).

Unless cultures are shaped to such an extent by their subsistence economies that nothing else signifies (which appears highly unlikely), the only way we can hope to learn what small-scale societies were like prior to contact with more complex ones is through archaeological research. Modern hunter–gatherer and small-scale agricultural societies do not provide reliable analogues. That means that the technique of conjectural history, which since the eighteenth century has privileged the use of ethnographic data for understanding social and cultural change, can no longer be relied on for that purpose. Social anthropology may turn out to be only a study of acculturation, and archaeology may provide the only reliable information about early stages in the development of human societies as well as the baselines for understanding changes brought about by European and other forms of contact. Hence archaeology, not social anthropology, may constitute the basic framework for anthropological research. Even if these claims have underestimated the capacity of small-scale societies to resist colonization and control by technologically more advanced groups, the issues raised are sufficiently fundamental that they require careful empirical study. The full impact of Western colonial expansion on most smaller societies can only be assessed archaeologically. However this particular issue is resolved, studies which consider interaction between societies as a source of change stress the complexity and multilinearity of sociocultural evolution to a much greater degree than do unilinear schemes.

It has often been argued that, because technological innovation has to build on existing knowledge, it must to a large degree follow a fixed order (Carneiro, 1970a; Harris, 1968: 232). Various studies have suggested that there is considerable cross-cultural regularity in the order in which various technological inventions have been made in different societies. Yet, as a result of early Chinese advances in producing heat-resistant ceramics, they were able to cast iron almost 1500 years earlier than were West Asian and European metallurgists, even though iron-working had begun in Western Asia much earlier than it had in China. Likewise, while agriculture generally preceded and may have led to the development of states and civilizations, chiefdoms based on intensive collecting have been found in various parts of the world, such as Florida; and in Peru monumental architecture preceded the development of agriculture (Burger and Salazar-Burger, 1993; Moseley, 1975; Widmer, 1988). Thus, even in the domain of technology, care must be taken not to read too much uniformity or linearity into cultural evolution. This was recognized long ago by Lubbock and Pitt-Rivers.

Neo-Weberian Critiques

Evolutionary linearity is further called into question by Weberian-inspired historical sociologists such as Michael Mann (1986) and Ernest Gellner (1988). Mann sees a general pattern to socio-cultural evolution, in the form of parallel developments occurring independently or nearly independently of one another in widely separated regions, as traceable only to the end of the chiefdom stage. He maintains that there are too few examples of pristine early civilizations or of later forms of complex societies to permit trends to be distinguished from singularities. This view echoes Karl Popper's (1957: 108–11) claim that true laws have to apply to many societies; otherwise they are merely descriptions of a 'unique historical process'. Mann's identification of only six civilizations that can be assumed to have developed independently of one another is very different from Yoffee's (1993: 71) assertion 'that states are not rare and precious entities in the evolution of human societies and do not require special [particularistic?] explanations . . . for their development'. Sanderson (1990: 218) maintains that six examples are sufficient to draw conclusions and sees parallel evolution characterizing the origin of early civilizations, although he denies that sociocultural evolution displays any single master trend. Both Mann and Gellner treat the birth of industrial civiliza-tion in Western Europe as the outcome of a highly complex conjunction of economic, political, social, and religious develop-ments that might never have occurred simultaneously anywhere else or at any other time in human history. The industrial revolu-tion is an event that Gellner (1988: 261) does not hesitate to label 'miraculous'. Hallpike (1986: 371) has adopted a similar position by claiming that capitalism is not a universal stage of human development but a unique expression ('oddity') of Indo-European cultural principles.

Gellner (1988: 213–23) viewed industrial societies as displaying many national and regional variants that he believed would persist and even grow more pronounced in the future. In the past other social scientists have preferred to distinguish two main variants of industrial society, capitalist and socialist, which they viewed as equally viable or at least as both able to survive for a long time (Sebag, 1964). Since the spectacular collapse of the Soviet Union and the abandonment of its planned economy, the New Right has been emphasizing the superiority of the capitalist variant and its status as the only form of industrial society that is viable in the long term (Fukuyama, 1992). Even if the origin of industrial societies is treated as a miracle, their later development is increasingly being viewed as a linear process.

By treating singularities and evolution as antithetical concepts, Mann identifies sociocultural evolution with developmental morphologies rather than with the underlying processes that bring about greater social complexity. Depending on the situation, such processes can result in singularities or parallel developments. Both Mann and Gellner imply that the combination of factors that created industrial civilization in Europe is too complex ever to occur again. At a specific level this argument has much to recommend it. Many of the key technological innovations that encouraged the economic development of Europe (harnesses for horses that permitted more efficient agricultural production, gunpowder, paper-making, movable type, and cast iron) were known in East Asia earlier than in Europe (White, 1962). Yet the impact that these innovations had on the rivalrous societies of Europe was much greater than their effect on China.

Yet, given how little we understand about the underlying processes of sociocultural evolution, the claim that industrial civilization as a generic type could have originated only in Europe seems unwarranted. As a result of its very nature, the influence of industrial civilization spread so quickly as to rule out any possibility of a second society of that sort developing independently. Even if during the last 800 years conditions were more propitious for the rise of capitalism and industrial civilization in Europe than they were anywhere else, that would not rule out the possibility that, if European civilization had failed to evolve an industrial economy, a civilization based on industrial production might not have developed somewhere else at a later date. Despite the various reasons that have been adduced why Chinese civilization would not have industrialized (Hall, 1986; Hallpike, 1986: 324), China has gone through periods of greater and lesser control by government bureaucracy and has experienced marked population fluctuations. It is conceivable that sometime in the future conditions would have been propitious for an industrial revolution to have occurred in China. The resulting industrial civilization would have been different in many respects from European industrial societies, because Chinese civilization is different, but it would have possessed many of the same economic, social, and ideological prerequisites and the same power to co-opt and transform less developed societies.

This raises the question of what impact specific cultural factors have on sociocultural change. It could be argued that Western civilization would not have evolved had it not inherited from ancient Rome the legal distinction between private and public property (*res privata* and *res publica*). Without the specific protection that the concept of private property provides for the individual accumulation of surpluses, it might have been far more difficult, or even impossible, for capitalists to build an industrial economy. Yet industrial economies have spread into countries where this distinction was unknown. Does this mean that capitalism and an industrial economy can thrive without this concept, or is its dissemination essential for successful industrialization? In addition, can the concept of private property effectively exist without a corresponding respect for the notion of public property? This is not an idle question since modern capitalists, in an effort to increase corporate profits, are seriously undermining respect for public property and seeking as far as possible to discredit the concept (Saul, 1995). We are dealing with a pair of concepts that appear to be important for the successful functioning of an industrial society. But it is unclear whether they were a prerequisite for such a society to develop or whether something like them would have evolved as part of the complex of institutions to which any industrial society gave rise. With only one example of a pristine industrial society, this is the sort of question that is difficult, if not impossible, to answer.

Recent critiques of neo-evolutionism by postmodernists, and by processual evolutionists such as Flannery and Marcus, have called into question the desirability of equating sociocultural evolution with unilinear change. Sociocultural evolution clearly involves more than a progression through stages of increasing complexity. Even the development of technology displays an unexpected degree of multilinearity. Progress is a subjective concept and today few people would accept a direct connection between technological and moral development. What appears most clearly to characterize changes in societies and cultures in general, and in many specific instances, is a tendency towards increasing complexity. This takes the form of increasing heterogeneity, with both societies and systems of knowledge coming to be made up of ever-greater numbers of horizontal components, which in turn require increasingly elaborate systems of hierarchical control to integrate them (McGuire, 1983). Individual societies suffer collapses and can move down, as well as up, a scale of complexity. Yet there is no evidence that at any one point of time in the past all, or even most, human societies have suffered such a relapse. Nor, when societies become less complex, do they revert to being the same kind of simpler society from which they originally developed. Some knowledge and behaviour patterns that have been acquired at a higher level survive the collapse and play a role in recreating complexity. Thus sociocultural development appears to have a general morphology, as well as an underlying process. That morphology is a tendency towards greater complexity rather than a more specific form of linearity.

Humans as Historical Agents

Postmodernism also challenges the ability of a neo-evolutionary approach to understand processes of change. It is argued that neither societies nor cultures, the principal units of neo-evolutionary analysis, are bounded entities. Nor, although they must provide for basic human needs and functional requirements, are sociocultural systems now thought to be as tightly integrated as many sociologists and social anthropologists once maintained. Networks of social relations and cultural patterns not only have complex distributions within societies but frequently cut across a number of different societies. This is held to render problematical the very idea of societies as valid units of analysis (Mann, 1986: 1–2; Yoffee, 1993). It is also pointed out that all human behaviour is guided by perceptions and that the more purely symbolic components of cultural systems are only subject to a limited degree to the same rigorous selection as are practical aspects of human ecological adaptations. Boyd and Richerson (1985: 274) suggest that the evolution of the purely symbolic aspects of culture can occur almost completely independently of functional considerations. It is argued that all these factors make any holistic understanding of sociocultural change far more difficult than neo-evolutionists have imagined (Kiser and Hechter, 1991; Sperber, 1985).

Postmodernists sensibly maintain that individuals, not societies or cultures, are social actors. They argue that it is necessary to start at the micro-, or personal, level in order to understand socio-cultural change, which occurs as the collective outcome of individual decisions. Change has to be understood in terms of the specific priorities that guide individual behaviour, or that of groups made up of like-minded individuals, even when an analysis is conducted largely in terms of the socially perceived costs and benefits of alternative courses of action (Shennan, 1993: 56). Anthropologists are currently investigating how changes in cultural priorities are brought about by defining new focal points to co-ordinate actions, competitive selection among social contracts, and bargaining (Ensminger and Knight, 1997).

Culture is more than an epiphenomenal mnemonic device for processing energy and adapting human groups to their environment. It is also knowledge inherited from the past through tradition and upbringing that endows perceptions with meaning and provides guidance for future action based on past experience. While such an inheritance can be modified or manipulated for personal or group ends, much of it is simply accepted without question and to that extent it constrains human behaviour. Henri Frankfort (1956) rightly or wrongly attributed the differences between ancient Egypt, a large state ruled from early times by a divine monarch, and Mesopotamia, which began as, and long remained, a collection of city states, to very different cosmologies that he believed remained constant throughout their history and were the mainsprings of their creativity. Hallpike (1986: 288) has argued that the core principles of society, which he defines as general beliefs about the cosmos and social values that can persist for millennia and are not functionally related to the mode of subsistence, provide guidance for organizing knowledge and social life, help to sustain cultural systems, and can resist change. Examples that he gives of core principles in Chinese civilization include a belief in unequal and opposite forces and entities forming wholes, the primary importance accorded to patrilinear descent, the importance of divination, the badness of competition and litigation, and the primacy of the concept of order over that of law (1986: 312–29). Such ideas, he maintains, are not simply epiphenomenal by-products of contemporary social life but a cultural inheritance that plays an independent and major role both in bringing about and resisting social change. If the decisions people make must be understood in terms of their perceptions of reality, understanding is required of the cultural as well as the ecological constraints that influence such perceptions. It is objected that neo-evolutionism ignored cultural factors that promoted and inhibited cultural change, especially at the band, tribal, and chiefdom levels.

While much of the postmodern critique has been directed against the limitations of neo-evolutionism, its preoccupation with the unique and the parochial makes it critical of any evolutionary perspective. Moran and Hides (1990: 212) have objected that postmodernism's rejection of supra-local problems precludes any investigation of such important issues as 'the origins of humans, human societies and civilisations in the ancient past'. Andrew Sherratt (1993: 125) argues that, by focusing its attention so exclusively on the study of isolated, specific cases, postmodern social science precludes an effective application to broader issues of its most valuable insights relating to the role and meaning of material culture and the creation of value. Ironically, Sherratt maintains, that in turn bars an approach that takes pride in its critical attitude from gaining 'any critical insights into the larger structures which underlie the growth of political hegemony and the emergence of capitalism'. These are important observations.

Evolution as Selection

Despite its great diversity and complexity, sociocultural evolution displays considerably more linearity than does biological evolution. Biological evolution follows a dendritic pattern. Species, once formed, are reproductively isolated from each other and can only persist, die out, or diversify to form new species. It is generally agreed that any attempt to define a central line or trunk leading to human beings would be hopelessly anthropocentric. One feature of biological evolution has been the development of increasingly complex forms of life. Even so, biologists such as Stephen J. Gould (1996) maintain that there is no trend in biological evolution that impels simple creatures toward complexity. If some later forms are more complex than earlier ones, that is because the first creatures had to be small and simple and thus any increase in variation had to be in the direction of greater complexity. Yet, if biological evolution lacks any obvious directionality, it is because species are reproductively isolated from one another and a myriad of ecological niches can accommodate a huge variety of different forms. Where simpler and more complex types compete for the same ecological niches, the more complex forms tend to replace their simpler rivals.

In the case of human beings, there is nothing analogous to a reproductive barrier separating one culture from another. Humans are by nature equipped to abandon the ways they are used to doing things and adopt new ones. Culture, in the generic sense of the ability to learn, thus becomes the primary niche to which all humans adjust. Under such conditions, conventional ecological niches provide limited and ever-diminishing protection for small-scale societies against the transforming and absorbing pressures of larger and more complex ones. As a result, sociocultural evolution acquires ever more directionality.

All human societies have evolved from hunter–gatherer and probably still earlier scavenging societies, none of which would have had more than a few hundred members. Since then human history has been characterized by the development of societies of ever greater scale and complexity. Beginning about 10,000 years ago, small-scale agricultural societies replaced hunter–gatherer ones over large areas of Eurasia, Africa, and the Americas, and these in turn gave rise in various regions to early civilizations. Today societies everywhere are being transformed or dismembered by a rapidly evolving industrial technology that began in Europe about 250 years ago. Because of the greater technological, economic, and military power of industrialized societies, all other societies have had to adjust in various ways to their demands and practices in order to survive. Small-scale non-industrial societies break down when they are unable to defend their land base or to feed growing populations. Larger non-industrial societies must industrialize or adjust themselves to supplying the raw materials needed by industrial ones. Less developed societies may resist change or struggle to preserve their identities in the new world order. Traditional values or belief systems may be enhanced when they help to promote economic modernization while at the same time reinforcing traditional identities, as Islam appears to do very effectively (Gellner, 1988: 216–17). In spite of this, it is impossible to explain what is happening to any part of the modern world outside the framework of a global industrial economy to which all societies have to adapt as best they may.

Throughout history, more complex societies have possessed more diverse and powerful technologies that have supported denser populations, integrated larger areas, and transported people and goods more effectively than do less complex ones. They have also possessed a more diversified social organization based on increasing specialization and have been able to co-ordinate activities on an ever larger scale. The growing mutual dependence of the members of such societies has imposed ever greater discipline on them. Their religious beliefs and secular values have encouraged greater self-discipline and commitment to carrying out goals defined by the society as a whole or by its leaders. In the language of biological evolution, complex societies manifest not only greater general complexity, organizational coherence, and an enhanced ability to deploy energy but also superior flexibility, self-regulation, communications potential, and adaptive skills.

A related characteristic of more complex societies is their enhanced capacity to influence weaker neighbouring ones. While the number of autonomous societies, as well as linguistic and cross-cultural variation, increased as long as small-scale societies continued to spread around the globe, the development of larger and more complex societies has resulted in a decline of such diversity, although this has been offset by increasing cultural variation within larger societies. Part of this change takes the form of dominant societies imposing their own political, economic, and social institutions on weaker groups. This has been a feature of European colonialism since the fifteenth century. Major structural impacts also include nations such as Japan selectively developing new political and economic institutions that have permitted them to survive and compete within the world system and small indigenous societies being encapsulated, if not absorbed, by new nations formed as a result of colonization.

The diminution of cross-cultural variation has come about not only as a result of the modification and suppression of technologically less-developed societies by, or in necessary response to, more complex ones; it also occurs as a result of individuals voluntarily abandoning familiar ways of life that no longer appear viable or attractive to them. The positive attractions of more complex societies take many forms. Some aspects of advanced technology obviously make life safer and easier. These include medicines that save and prolong life and diminish pain and suffering, as well as numerous labour-saving devices. Consciousness-expanding technologies, most notably cinema and television, provide welcome diversion in the third world and expose people to novel attitudes and forms of behaviour that they may judge to be improvements by comparison with traditional ways of doing things. Finally the desire to identify with the rich and the powerful leads people to seek to adopt the material culture of more complex societies as status symbols. Julian Huxley (1960: 77–8) has argued that more complex societies offer many human beings enhanced opportunities to realize their individual potential. Sanderson (1990: 49) notes that, in general, cultural diffusion exhibits selective directionality: people living in less complex societies are far more likely to adopt ideas from more complex ones than those living in more complex societies are to borrow ideas from less complex ones. This is not to claim that people living in more complex societies cannot, and do not, learn from smaller ones. This happened frequently in the early stages of European colonization (Trigger, 1985). Rather it signifies that attitudes, power relations, and the attractions of seemingly more successful lifestyles tend to channel emulation in their direction.

It is also not true that the quality of life in simpler societies is inevitably improved by contact and emulation. An isolated, small-scale society may be economically poor but enjoy a satisfactory community life. On the other hand, societies being drawn into larger economic systems tend to be rent by inequalities and many of their people degraded by exploitation. Hence, even if the overall standard of living were to rise, the quality of life for most people might decline (Baum, 1996: 43). But, by the time that is realized, people living in smaller societies may have forgotten vital parts of the knowledge that once permitted them to be economically independent, they may have increased in numbers beyond the carrying capacity of their environment, and have come to rely for their survival on tools or goods they cannot produce. Thus they become locked into a dependent relationship.

It has also long been recognized that over the course of human history the development of greater sociocultural complexity and the impact of such complexity on neighbouring societies have been proceeding at an accelerating rate. The tempo of innovation increases as more knowledge accumulates and more efficient techniques for securing and processing data are developed (Huxley,

1960: 225). This results in more complex societies changing ever faster and being able to cope more effectively with new situations as they arise. As societies grow more complex, they are able to dominate less developed neighbours ever more quickly, thoroughly, and completely. Huxley (1960: 25) has suggested that new types of sociocultural organization are like biologically dominant species that spread at the expense of earlier forms. Hallpike (1986: 75–6) has observed that where competition is weak, inefficiency can last forever. Yet, where societies must confront more complexly adapted and powerful neighbours, relative weakness can quickly become a lethal defect. Hallpike also suggests that this explains why there is so much more cross-cultural variation among small-scale societies than among larger ones. As societies become larger and competition increases in intensity and tempo, the opportunities for less fit systems to survive diminish rapidly. As a result, sociocultural evolution acquires greater overall direction as societies become more interactive on a regional and eventually on the global level.

Social and cultural changes are brought about by different mechanisms. Cultural change is mainly the result of personal choice, which involves weighing the attractions of novelty against those of familiar ways of doing things. Structural change occurs as a result of the capacity of more complex and powerful societies to dominate and encapsulate less efficient ones. As societies grow more complex these two processes become increasingly interconnected. Today, by means of advertising and popular culture, advanced societies are able to use the mass media to try to predispose people in less developed countries to aspire to possess the material trappings of a Western lifestyle, and are often very successful at doing this. This in turn becomes a force encouraging extensive economic and social changes in such societies (Potter and Welch, 1996).

Sociocultural evolution is different from biological evolution at the species level because cultural behaviour is learned and therefore can be altered as a result of personal choice or compulsion. This makes it much easier for the ways of life associated with more complex societies to spread at the expense of those of less complex ones. Unlike a biological species, a more complex culture does not have to spread by replacing the individual members of a less complex society with its own people. More often it expands by undermining the viability of the weaker culture and persuading or compelling people to adopt a new style of life. In this way new beliefs and forms of behaviour spread among existing populations.

Postmodernists, who value cultural diversity for its own sake, disapprove of such developments, which they see as dangerously homogenizing the planet. They emphasize approvingly examples of the tenacious survival of ethnic identities and of traditional cultures despite colonial oppression. They also try to minimize the extent to which smaller-scale societies have been integrated into larger political and economic units (Freidel et al., 1993). In effect they seek to protect other people from the dynamism and change that for better or worse has made their own way of life and their scholarship possible.

Yet, despite what anyone may wish to have been the case, because of the relative ease with which human beings are able to modify their behaviour, linear change has played a more prominent role in human history, and radiation along divergent lines of descent a correspondingly weaker role, than has been the case with biological evolution (Ingold, 1986: 293). Hence, the development of successively more complex levels of social organization and of culture is a more prominent feature of sociocultural evolution than is the development of higher levels of complexity in biological evolution. As a result of their preoccupation with the unique and the idiosyncratic, postmodernists and neo-Weberians, like the Boasian and culture-historical anthropologists of an earlier era, have failed to note that human history does have a linear directionality. While this shape may not be as uniform as neo-evolutionists once believed, it provides important insights into some of the most crucial processes that have influenced human behaviour. While the concept of general evolution has served to celebrate and justify Western accomplishments, it also sheds light on the nature and motivations of colonialism, economic exploitation, and other forms of oppression, which postmodernism ubiquitously deplores but does not seem equipped to understand.

This approach also provides an explanation for the overall linearity of human history that, despite the considerable differences that have been documented between biological and sociocultural evolution, generally accords with evolutionary theorizing in biology. Linearity is no longer seen as a form of progress that is somehow built into the natural order, as the deist philosophers of the Enlightenment and most nineteenth-century evolutionists believed. Nor is it the product of some equally mystical long-term human striving to control nature more effectively and make human life easier and more secure. Human behaviour *is* purposeful but individual goals tend to be short term. As more complex societies arise, they create conditions that make the survival of neighbouring, less complex cultures increasingly difficult.

Until recently, hunter–gatherer societies were able to survive in remote desert and northern environments that were not suitable for pastoral or agricultural exploitation. Today, however, an increasingly effective system of transportation and the growing demands of industrial economies for raw materials have drawn even these regions and their peoples into the orbit of the world economy. The cultural equivalent of natural selection produces unilinear trends as a result of more complex societies seizing control of the resource bases and undermining the cultural distinctiveness of less complex ones. As cultural complexity increases, the power of the most advanced societies to do this is enhanced. This does not explain why societies become more complex, but it ensures that, even if only a very small number do, this will push increasingly large segments of humanity in the direction of greater sociocultural complexity. Hence even historical accidents that resulted in greater complexity could shape the general course of human history. Directionality results from what kinds of sociocultural systems survive and do not survive as a result of this sort of competition.

As is the case with biological evolution, survivability has to be measured in the short term. If industrial civilization turns out to be autodestructive as a result of the damage it is inflicting on the world ecosystem, it could be argued that in the long run it would have been better had all human societies remained at the hunter–gatherer level. If culture has provided human beings with a form of adaptation that has enabled them so far to expand more rapidly and exert more influence on the ecosystem than any other species, there is no assurance that in the long run this adaptation will not prove maladaptive or flawed. Yet this ceases to be a proposition with any evolutionary significance once it is acknowledged that the evolution of intensive collecting societies or village cultivators created short-term selective benefits, perhaps mainly in the form of larger populations, that spelled the demise of most neighbouring hunter–gatherer societies once the two types of societies began to compete for the same resources. It is not what happens in the long run, but in the short term, that determines the course of evolution.

This view of sociocultural evolution is essentially similar to Leslie White's (1949) theory of general evolution. White argued in a less nuanced fashion that, because the most advanced culture of any era spelled the eventual demise of all less developed cultures, it was only necessary to study a linear sequence of successively more advanced cultures. While this view, as Sahlins and Service (1960) and many others realized, rendered most of human history irrele-vant to an unacceptable degree, White had succeeded in identifying the major factor that was producing linearity in cultural evolution. This view accords with Marvin Harris's observation that parallel and convergent evolution play a much more prominent role in sociocultural evolution than does divergent evolution (Sanderson, 1990: 155) and Hallpike's (1986: 211) assertion that orthogenesis is normal in social evolution.

Culture and Adaptation

Yet, if social competition tends to promote linearity, culture itself produces diversity. Few anthropologists accept the claim that culture is an epiphenomenon shaped in all its details by the manner in which individual societies adapt to their environments. Cultures are above all collections of knowledge transmitted from the past which provide individuals with a guide for daily living, familiar patterns on the basis of which innovations can be attempted, and a set of values that to some degree can inhibit or channel the direction of change. The main disagreements centre on the extent to which cultural patterns influence social behaviour and these patterns are resistant to change (Welch, 1996a). Structuralists view culture as having a formal coherence and unity that imposes significant limitations on the thought patterns of anyone raised in a particular culture. They therefore tend to view cultural patterns as highly resistant to change, and this has led some of them to reassert the uniqueness and incomparability of different cultures (Leach, 1970). At the other end of this spectrum, ecological determinists and Marxists emphasize the relative ease of social change as a result of rational calculations of efficiency or the struggles of different interest groups to control the resources of societies for their own benefit.

Michael Mann (1986) discusses four sources of social power – economic, political, religious, and military power – and has tried to specify the different ways in which these powers influence human behaviour. He distinguishes between intensive power, which relates to strongly centralized control over smaller, tightly integrated areas, and extensive power, which influences behaviour over larger areas that lack political integration. For example, religion and trade can often promote interdependence and shared values on a much larger scale than can military activity. In the Weberian tradition, however, Mann does not try to rank these powers. Maurice Godelier (1977: 63–9; 1978) argues that relations of production can be rooted in any part of a sociocultural system, including its religious organization, but maintains that the success of capitalist societies has to do with their ruthless disembedding of the economy from other social institutions. While these varied positions have stimulated vigorous debates, social scientists have so far failed to agree on a detailed body of theory that would explain human behaviour to the satisfaction of even a substantial minority of scholars. Childe's view that a culture's understanding of the natural world has to correspond sufficiently with that reality for humans to be able to act on it effectively precludes a successful culture from being merely anything that its members individually and collectively wish it to be. Yet, because they are complex, symbolically mediated systems, cultures also create cosmologies and concepts of good and evil that on the one hand can disguise self-interest as altruism and on the other mobilize support against self-interested behaviour that threatens the welfare of the collectivity. By its very nature, culture provides great opportunities for deception and self-deception, but also for more far-sighted and objective understanding.

Especially in the more purely symbolic realm, many diverse beliefs or practices may be equally efficacious. Because of this, the cultural content of societies at similar levels of development differs radically and even cultures that stem from a common ancestral one may be quite different as a result of each producing its own innovations that make sense from an internal, structuralist point of view but have little, if any, direct adaptive value. Yet, because culture influences human behaviour, it has a constraining power analogous to that of the natural environment, although different in the way it operates. While the natural environment constrains as a result of the necessity for humans to allocate scarce resources among competing ends, culture constrains through the need to impose some sort of cognitive and psychological order on a potentially unlimited production of ideas. Karl Marx was the first to understand the complex relations that exist between economic constraints and culturally mediated human behaviour when he balanced his claim that 'it is not the consciousness of humans that determines their existence; it is on the contrary their social existence that determines their consciousness' (Marx [1859] in Marx and Engels, 1962, 1: 362–3) with the statement that 'human beings make their own history . . . not under conditions chosen by themselves, but ones directly encountered, given, and transmitted from the past' (Marx [1852] in Marx and Engels, 1962, 1: 247). Similar views about the cross-cutting relations between culture and economy, although not all of them are grounded in a materialist epistemology, have characterized the work of Hallpike (1986), Gellner (1988), and others.

There is thus an urgent need to study empirically the kinds of roles that are played by all sorts of social and cultural factors and to understand their relative importance and how they relate to one another. It is possible to document a wide range of differences in key areas of cultural values and economic organization in societies that are at the same general level of complexity. Contrary to what Julian Steward assumed, early civilizations display great variation in agricultural technology, population density, patterns of land ownership, techniques for enforcing and appropriating surplus food production, and the ways in which the manufacturing and distribution of craft production were organized (Trigger, 1993: 27–44; for a modern example see Welch, 1996b). Even if some forms of organization appear to have been more effective than others, these differences were not of great importance, provided that these societies were not competing with better-organized rivals.

Wide differences are also apparent in the social values of early civilizations, especially as these relate to elite male roles. Different early civilizations viewed the ideal life for a man very differently. The Egyptians admired the well-trained and emotionally self-disciplined bureaucrat; the Aztecs the courageous and valiant warrior; and the ancient Mesopotamians the wealthy landowner. Like art styles, these ideals seem to have been idiosyncratic to specific civilizations. Yet, unlike art styles, they influenced behaviour throughout the whole society. The highland Mesoamerican idealization of the warrior was part of a theological complex in which war and the shedding of human blood were viewed as essential to sustain the universe. These beliefs in turn exerted vast influence over interstate relations. The idealization of the bureaucrat may help to explain the chronic problems of military leadership that bedevilled Egyptian relations with their more bellicose neighbours (Trigger, 1993: 84–5; see also Hall, 1986).

Yet early civilizations shared important general features. All of them exhibited well-developed social hierarchies and economic inequality. In each civilization a very large portion of the agricultural surplus fell into the hands of a small privileged group. The ways in which this surplus was extracted from its producers were highly variable. It was collected as taxes, tribute, land rent, revenues from sharecropping, and in other ways. But the result was always the same. These surpluses ultimately supported an upper class whose interests were defended by the power of the state. An Egyptian bureaucrat or an Aztec military officer was as interested in preserving his economic and social privileges as was a Mesopotamian landowner, and all three were prepared to use the coercive powers of the state to do so (Trigger, 1993: 44–54).

There was also surprising cross-cultural uniformity in religious concepts. Despite a vast diversity in specific religious beliefs, those who lived in the early civilizations appear to have agreed that supernatural forces provided the energy that kept the universe functioning and reproducing itself, and that human beings in turn provided the energy that sustained these supernatural forces. This energy was transmitted to the supernatural realms in the form of offerings, mainly made to the gods by the king and members of the upper class. This set of beliefs appears to have been independently replicated in the various early civilizations as a result of the universe being conceptualized as a projection into the cosmic realm of the unequal economic arrangements that prevailed in these civilizations. The peasants physically supported the upper classes, who in turn ensured a harmonious political order without which an agricultural economy could not have produced enough food to support either the peasants or the upper classes. By extrapolation it was believed that, while the deities were essential to maintain the cosmic order, without peasants to produce the food needed to sustain the gods the cosmic order would perish. The king and more broadly the ruling classes served as intermediaries in the flow of energy between the human world and that of the supernatural (Trigger, 1993: 94–105).

These beliefs defined essential roles for ordinary people, rulers, and supernatural forces in the operation of a mutually beneficial cosmic order and in the course of doing so endowed each social grouping with their own rights and duties within that order. That in turn served to check the abuse of power by the upper classes. It was possible for ordinary people or subordinates to criticize a greedy or incompetent ruler on the grounds that he was failing to do his duty to the gods, and hence for failing to maintain the cosmic order, without challenging the claim that the king was above all other human beings and therefore answerable to no other mortal. The early civilizations were societies in which religion had replaced kinship as a medium of political debate. The similarities in the fundamental structure of religious beliefs in the early civilizations and the constitutional role they played suggest that they were not simply random ruminations or haphazard refinements of earlier belief systems, but a projection of the key features of early class societies into the realm of the supernatural.

Biological Universals

Economic, social, and cultural universals, such as are manifested in the similar patterns of behaviour and beliefs that evolved in the early civilizations, appear to be expressions of a shared human nature (Brown, 1991). Hence biological evolution may explain much more about human behaviour than does discursive philosophy. The higher primates who are most closely related to human beings are all social animals. They live in groups and support and depend on each other in various ways, but also tend to be hierarchical and competitive. It is generally assumed that the need for co-operation within human groups has been enhanced by greater reliance on technology, more complex forms of learned behaviour, and the increasing role played by symbolic mediation (Carrithers, 1992). It is clear, however, that these developments have not totally suppressed competition within human groups, including nuclear families.

It is generally recognized that biological evolutionary concepts such as natural selection and the survival of the fittest cannot be used to rationalize all forms of human behaviour, including war, slavery, economic inequality, and neo-colonial economic exploitation. Yet, between the 1940s and 1960s, Raymond Dart (1949) and Robert Ardrey (1961, 1966) sought to popularize the idea that millions of years as hunter–gatherers had enhanced hominid cruelty and bloodlust and turned modern human beings into 'killer apes'. More recent works have argued that this common experience of all human ancestors has placed a premium on male aggressiveness and equipped humans to respond well to short-, but not long-, term challenges (Simmons, 1993: 171–2; Wrangham and Peterson, 1996). Marxists, on the other hand, have maintained that there is no universal human nature, only various natures determined by specific relations of production (Bloch, 1983: 80). Marx and Engels looked to the sharing and co-operation found in hunter-gatherer and small agricultural societies as evidence that greed, competitiveness, and self-interest were not inevitable characteristics of human behaviour. They also believed that these societies provided evidence of what human behaviour might again be like once technologically advanced societies based on private ownership of the means of production had been replaced by societies based on communal ownership. Lacking adequate ethnographic documentation, Marx and Engels failed to realize that in small-scale societies laziness, stinginess, and self-assertion existed but were severely discouraged by ridicule, gossip, ostracism, and fear of being accused of witchcraft or of falling victim to it as a consequence of antisocial behaviour. On the other hand, hard work, generosity, and modesty were rewarded by favourable public opinion. In these societies selfishness and self-assertion were opposed by the power of public opinion, just as in state societies private property and the authority of leaders are protected by state-sanctioned violence. Yet public opinion is an effective curb on behaviour only in small social units, where everyone knows everyone else, and in situations where geographical mobility permits dissident groups to strike out on their own rather than having to submit to coercion by other factions within their society (Lee, 1990; Trigger, 1990).

Marxists and economic substantivists such as Karl Polanyi likewise claim that marketing behaviour developed only recently (Polanyi, 1944; Polanyi et al., 1957). They oppose economic formalists, who treat the profit motive as a universal characteristic of human behaviour. Yet examples of profiting from economic transactions, as well as of reciprocity and economic altruism, can be documented at all stages of social evolution. The extent to which each of these relations is practised is determined by the degree to which the economy is embedded in other institutions, such as the state, religious cults, or the family.

One of the principal weaknesses of Enlightenment philosophy and of Marxism has been their belief that human beings are naturally inclined to be reasonable, generous, and co-operative. Marxists have also believed that, if people could only be rescued from the corrupting influences of an exploitative class society, these positive qualities would automatically manifest themselves. In the Soviet Union this naively optimistic ideology, combined with a lack of public accountability by those who controlled the state, created fertile conditions for the unchecked spread of corruption, which ultimately destroyed the ideals professed during the October Revolution and later by the Soviet state. Christianity, with its recognition of the selfish and cruel as well as the benevolent components in human behaviour, provides a far more realistic analysis of the human nature that has resulted from millions of years of biological evolution, even though that analysis has been expressed in the language of religious myths.

Our primate ancestors were social creatures before they became capable of cultural behaviour. Such behaviour, which involved enhanced memory, novel forms of symbolic manipulation, and greatly expanded communicative skills, permitted greater behavioural flexibility, and that in turn allowed humans to adjust to an increasing variety of ecological niches. Yet viewing culture as an extrasomatic means of adaptation does not necessarily involve a commitment to ecological determinism. Symbolically based behaviour, because of its reflexivity and self-reflexivity, had the capacity both to transform existing primate needs and to create new ones. This produced what Dan Sperber (1985: 80–1) calls 'side effects' of biological evolution. According to Sperber, much symbolic manipulation affects human well-being only marginally, and hence does not come under as strong selective pressures as do activities related to food production. Thus there is wide variation in the myths, legends, religious beliefs, and art styles associated with different cultures. In this, he agrees with Boyd and Richerson (1985: 272–9).

Other 'side effects' are, however, less trivial. The most striking of these is the ability to conceptualize alternative courses of action and choose between them. This endows individual and collective behaviour with a moral dimension. While morality is culturally defined and variable, it is supplied with a pan-human dimension by the human capacity to see others as being like themselves and to imagine how others may view them. This creates at least the possibility of a society in which people enhance themselves by being open to the needs of all their fellow human beings (Baum, 1996: 76). Because of this capacity, far more uniformity may underlie myths, legends, and religious beliefs than many analysts would admit.

If culture is transmitted from the past while being modified to meet current needs, and if it possesses a moral dimension, it becomes possible to view human history without succumbing to the interpretative errors of ecological or cultural determinism. While not endowing human beings with omniscience, conscious-ness, choice, and planning play significant roles in human history and there is truth to Marx's claim that 'human beings make their own history'. On the other hand, much change comes about as a result of people imitating the novel fashions and behaviour of those who are regarded as successful. While this sort of behaviour is not particularly dangerous in a society that is changing slowly and where the practical effects of innovations can be monitored fairly easily, it can result in dysfunctional and highly dangerous occur-rences in more complex and rapidly changing societies (Boyd and Richerson, 1985: 277–9). This, of course, is increasingly the problem faced by modern industrial societies. If the trajectory of sociocultural evolution is shaped blindly at the highest level as a result of societies with greater selective power and flexibility replac-ing those that have less, but the dominant societies have reduced ability to monitor the long-term effects of social change, there is ever greater risk of sociocultural evolution becoming dangerously dysfunctional and self-destructive.

10
The Planning Imperative

The positivist contention that explanation and prediction are equivalent does not apply in the historical disciplines, whether they be cosmology, historical geology, palaeontology, archaeology, or human history. This is not because important generalizations do not exist that explain changes in these fields, but because the factors that influence particular events tend to be so complex that it is impossible to predict the future in any significant detail. A number of alternative developments generally seem possible. Yet, once the course of history has been established, it is possible to explain what has happened with reference to significant generalizations (Bunge, 1996; Roberts, 1996). This applies equally in the social sciences, where theorization is generally agreed to be weak, and in biology, which has long possessed a general consensus about the body of theory accounting for change. With the information about what happened provided by the palaeontological record, it is possible for biologists to explain how humans or horses have evolved. Yet, knowing only what existed in the early tertiary period, it would be impossible to predict that these, or any other particular species, would have developed.

Also, while human nature changes only very slowly, and while the archaeological record reveals that the modern world has been produced by an ongoing historical process that links it to the remotest past, the present situation in which human beings find themselves is radically different in many respects from even the recent past. Many of the most difficult problems humans are encountering are very different from any that their ancestors confronted in previous stages of cultural development. For millennia individual human groups have confronted problems of overpopulation, local resource depletion, environmental degradation, and hostile neighbours. Yet humans have never before had to cope with global problems arising as a result of economic interdependence, environmental pollution, depletion of natural resources, and threats of nuclear, chemical, and biological warfare. Can knowing what happened in the past, different as it was from the present, teach us how to deal more effectively with current problems?

This question is related to whether the course of sociocultural evolution is determined independently of human volition by environmental, ecological, biological, psychological, social, or cultural constraints. Does sociocultural change manifest preordained laws of history as vulgar Marxists and ecological determinists maintain? Ideas of inevitability were espoused by the French bourgeoisie in the eighteenth century in an effort to empower themselves in their struggle against absolutism, and were cultivated later by socialist revolutionaries in their battles with the industrial middle class. It is also expedient for those in power to ascribe policy failures, or unpopular policies, to natural causes or supposed historical laws rather than attributing them to their own greed, miscalculations, and folly. If the course of history is determined by forces that lie beyond the control of human beings, all humans can do is try to understand these forces and accommodate to their operation, as the physiocrats advocated in the eighteenth century. But if the complexity of the forces that bring about sociocultural change makes it unlikely that the course of human history is predetermined, then decision-making has a more positive role to play and knowledge and moral responsibility have major selective value. There is good reason to believe that choice plays a significant role not only in our daily lives but in shaping long-range trends in human history (Franklin, 1990). Moreover, as long as we remain uncertain about precisely how important a role choice plays in human affairs, it is in our interest to behave as if it plays a significant one.

It is argued that human beings seem best adapted by natural selection for making self-interested and short-term decisions, which relate most clearly to the needs of individuals and small groups (Boyd and Richerson, 1985). Yet during the last few millennia they have demonstrated the ability to apply such skills to managing the affairs of larger groups. This has required individuals and groups to co-operate and organize activities on a larger scale and to address problems that require more complex management over longer periods of time.

Under these circumstances, the best use that can be made of an understanding of the past is to suggest the direction in which current processes are moving and what the outcomes of strategies for dealing with analogous problems have been in the past. Posing questions at various levels of generality makes it possible to draw on experiences from longer or shorter segments of the past. That the present is dominated by a neo-conservative ideology that is a revival of a nineteenth-century liberal one rejected in the 1930s makes the period from 1850 to 1940 an especially fertile source of information that can be used to estimate the potential consequences of current courses of action. Other insights can only be drawn from much longer perspectives, including the entire course of human history.

A New Order?

Prior to modern times most long-distance trade was limited to luxury goods, which alone repaid the high costs of transportation. Following the industrial revolution, growing population densities in industrialized societies encouraged the importation of basic foodstuffs from vast distances, extending a division of labour to the entire planet. The electronic communications and data-processing revolution that has occurred since the Second World War has created a still more elaborate international economy that has undermined the power of national states to control their own affairs (Horsman and Marshall, 1994). It is reported that 46 of the world's 100 largest economies are now transnational corporations and that they control nearly two trillion dollars, the equivalent of the entire economies of Germany and France and two-thirds that of Japan (Ford, 1996). The impact of these changes has been examined in detail by Patricia Marchak in her book *The Integrated Circus: The New Right and the Restructuring of Global Markets* (1991), and by Robert Harvey in *The Return of the Strong* (1995). The growing ease with which investment capital can now be transferred from countries where labour costs are high to localities that offer more favourable investment opportunities permits transnational corporations to dictate political programmes even to the most powerful nation states. These companies seek to undermine the power of trade unions and to dismantle welfare systems in the developed countries and prevent their development in the third world. In this way they have succeeded in reducing taxes, especially on corporations, and in increasing their own profits at public expense.

This new economic order is promoted by a neo-conservative ideology that denounces government interference in the economy as wasteful and stultifying, at the same time as transnational corporations play governments off against one another to secure public subsidies and tax reductions. Public debts are blamed on the wastefulness of the welfare state and on the inefficiency and high costs of running government bureaucracies, not on lower taxes and higher earnings by corporations. The increasing cost of medical care brought about by more elaborate and effective forms of treatment, by deregulated drug prices, and by the Aids epidemic is used to justify dismantling medical aid programmes. It is alleged that personal relations between patients and doctors are more cost-effective and of superior efficacy to ones involving government participation, and that chronically sick and disabled people are best cared for by relatives and friends. It is also maintained that welfare (unlike unemployment) undermines initiative and self-reliance to the moral detriment of the individual. Neo-conservatism is opposed to any form of government intervention in the economy (except for low corporate tax rates and other forms of government subsidies) and maintains that unhindered individual initiative will promote economic growth and enhance personal self-reliance and self-esteem everywhere in the world (Brenner, 1994; Friedman, 1962; Hayek, 1960). Neo-conservatism is in fact a revival in a little-altered form of nineteenth-century liberalism. Being unchanged, it is new only in a chronological, not in an intellectual, sense.

Aided by this formidable combination of new technology and ideological support, the international economy has now broken free of the control of even the largest and most powerful states and can do worldwide what industrial capitalists did inside various countries during the nineteenth century. Currency speculation destabilizes economies and can be used to blackmail governments into compliance with the demands of transnationals. Widespread unemployment, problems in pursuing long-term economic policies, growing wealth disparities within and between countries, and the rapid accumulation of wealth in the hands of a minority of the population are characteristics of the new world order. Some of these disparities are offset by the illegal narcotics trade, but this further undermines law and order and creates its own extremes of wealth and immiseration within both rich and poor countries. Even if the total wealth of the human species is increasing, specialization and urbanization mean fewer people are able to buffer themselves against unemployment by producing at least some of their own food and other necessities, as many could do in the past. This creates a depressed underclass which varies in size from one country to another. It also creates large numbers of people who are fearful that they will not be able to hold on to their jobs or small businesses as the economy is wracked by cut-throat competition and major structural changes. Such developments obliterate the hopes for a secure lifestyle that most middle-class people in developed countries enjoyed in the recent past. This heightens personal psychological stress levels and encourages the development of irrational social phobias.

There is also growing concern about an alleged breakdown in law and order, which fuels neo-conservative demands for harsher and more repressive treatment of criminals. Growing fear of the victims of the new social order leads the rich and the powerful to seek protection from them by resorting to private protection agencies and isolating their residences, offices, schools, hospitals, and places of enjoyment from public access. This self-imposed physical isolation further increases the social isolation of the rich and powerful from the rest of society and heightens their social irresponsibility by making it less necessary for them to consider the negative impact that their behaviour is having on society as a whole.

Neo-conservatives tend to blame social problems on the decline of traditional values, especially those of the two-parent family. What is not acknowledged is that the traditional family has fallen victim to declining personal income levels and a rising consumerism that has driven most adults to seek full-time, paid employment. Rising expectations, the declining value of individual incomes, and growing belief that self-fulfilment comes about through having a successful career and lots of material possessions have undermined both the traditional family and the self-esteem of many people who are discovering that the reality of their lives does not measure up to their expectations (Tice, 1996). Despite the obviously disruptive effects of neo-conservative economic policies, the neo-conservative ideology, with its promise of long-term trickle-down benefits for everyone, its constant criticism of government waste, its emphasis on the crucial importance of personal responsibility, and its growing monopolistic control of the mass media, has been extremely effective in rendering criticisms of what is currently happening ineffectual.

These developments conform in a general fashion to a pattern of economic growth that Immanuel Wallerstein (1974) has traced since the rise of capitalism in northern Italy in the late medieval period. As capitalism spreads, it reduces its investments in regions where production is well established and labour costs have risen, in favour of new areas where labour is cheaper. Today's investors are looking to the emerging Pacific Rim countries and beyond for high returns on their investments, while insisting that workers in North America and Europe must be more productive, flexible, and demand lower wages and benefits than they have been receiving if they hope to compete. Marxists predicted that the growing contradictions between capitalists and workers would lead to increasing class struggle and eventually to revolution. This prediction was not realized, in part as the result of a vast, and among Marxists unexpected (Lenin, [1917] 1939: 62–3), increase in the efficiency of agricultural production, which has kept food prices relatively low and permitted many people in the developed countries to share in the prosperity that followed the Second World War. At the same time, the state, by enforcing minimum wages and supplying unemployment benefits, medical care, and old-age pensions, blunted some of the worst abuses of capitalism and lessened class conflict.

Since the collapse of the Soviet Union, there has been a growing tendency to abandon these controls and a widespread insistence that capitalism in its present form marks the 'end of history' (Fukuyama, 1992). The latter concept does not accord with observations that technological growth alters societies and values and that a successful capitalist economy must be a dynamic and constantly expanding one. Capitalism is the first economic formation in human history that has explicitly glorified change rather than stability as a normal condition and actively sought to bring change about. Pre-industrial commercial capitalism gave way in the late eighteenth and early nineteenth centuries to the entrepreneurial capitalism of the owner-managed factory. That in turn was superseded in the early twentieth century by corporate capitalism, as companies grew larger and more bureaucratically managed, and most recently by the transnational financial capitalism of the electronic age. Each new form of capitalism represented something different from all preceding forms and impacted negatively on the fortunes of previously successful entrepreneurs who were unable to adapt to new conditions. The dynamism of capitalism is such that, by penetrating ever more deeply into the third world and transforming it as well as the more developed countries in its own interest, it can continue to expand and evolve for a long time. Under capitalism, social, political, and economic change seems inevitable, even if it is impossible to predict precisely what forms future changes will take. This is hardly an 'end of history'.

To talk about capitalism as an 'end of history', in the more general sense of claiming that free enterprise societies will never evolve into ones based on significantly different economic principles as capitalism's capacity for expansion is exhausted, seems premature. That sort of discourse reflects the desire of neo-conservatives to assert the transcendence of their values and approach to life, even though such an attitude is totally out of keeping with the transformationary ideals of the system they support.

This, of course, is an outlook long shared by the more dogmatic Marxists, who saw sociocultural evolution culminating in an alternative end of history: the workers' paradise. Both extremes view evolution as a process leading to an inevitable and already-known goal and thus deny either chance or choice a role in shaping human history. Each sees its preferred goal as a changeless perfection, which is the antithesis of a view that equates evolution with ceaseless transformation. Each end-point is an expression of vested interest, although for Marxists this remained a goal yet to be achieved, while for neo-conservatives it is an arrangement to be defended. Given the persisting dynamism of capitalism, it is necessary to consider whether transnational financial capitalism and its neo-conservative ideology represent a trend that will continue for a long time, a rapidly passing aberration that will soon be corrected as stronger international mechanisms of political control are forged to supplement or replace inadequate national ones, or part of an ongoing cycle within capitalism in which the power of economic and political forces will continue to vary in relation to each other.

Nation states, which nurtured the development of modern industrial societies, seem singularly unable to cope with recent developments. The experiences of the twentieth century suggest that in certain respects big government is as inefficient as its critics charge. Bureaucracies tend to become self-serving and over time their administrators and employees put their own interests ahead of those of the people and the society they serve. At least since the writings of the Arab historian Ibn Khaldun in the fourteenth century, government diversions of resources that might otherwise have promoted economic growth have been seen as a major cause of the decline of states (Rosenthal, 1967). Still earlier Arab sources attributed to the Persian monarch Chosroe I the observation that a kingdom could prosper only if its ruler had the skill to curb the rapacity of his officials so that industry and agriculture might flourish (Adams, 1965: 71). Joseph Tainter (1988) has argued that the tendency for administrative machinery to absorb ever more resources has played a major role in weakening the economy and bringing about the collapse of every major state that has ever existed, and may continue to do so in the future. These tendencies are especially severe in societies where totalitarian regimes are able to suppress criticism or where self-serving institutions can flourish unchecked to the detriment of the society as a whole.

Yet transnationals pose even greater economic and social problems to the societies in which they operate. Modern conglomerates, unlike nineteenth-century companies, are usually run by managing directors who lack expert knowledge of the actual processes of production. Their main concern is to secure the highest possible short-term profits for shareholders and huge salaries for themselves through the buying and selling of companies and by asset-stripping. The production of high-quality goods to please customers is at best a means to an end, not a source of pride or a necessity for staying in business. Yet these conglomerates have the financial resources to drive smaller, more specialized, and customer-oriented businesses out of production. In recent years efforts have been made to curb bureaucracy and increase profits by contracting out major aspects of production, marketing, and administration. Yet, even if the upfront costs of such services are lower, these arrangements result in fewer workers receiving pension or unemployment benefits. The long-term costs of looking after such people create an additional hidden debt that is being imposed on future generations of taxpayers.

This ought to confront governments with some interesting policy questions. Should large corporations be broken up to prevent the growth of vertical or horizontal monopolies, as antitrust laws in the United States have attempted to do since the 1890s? Might companies be prevented from acquiring controlling interests in businesses that are unrelated to their special sphere of competence? Ought contributions to social welfare funds be required, as is done in France, for all workers, whether full-time or part-time, employees or self-employed? Today's governments are too weak economically and too much played off against each other by transnational companies to tackle such problems. In developed countries the public debt is blamed on excessive and debilitating government spending on social welfare. Many third-world countries have been trapped by international debts at least nominally incurred for purposes of development, and the smaller ones frequently find themselves caught in networks of rapidly shifting commodity prices that they cannot control. The problems of third-world countries are made worse when transnational companies, often after inflicting serious social and economic damage in their search for quick profits, abandon them to cope with these problems on their own once the exploitation of their resources is no longer deemed profitable (Levitt and McIntyre, 1967). One of the frequent responses to such exploitation and growing social inequality is the development of more xenophobic forms of nationalism. Nationalism may help to unite troubled societies but it can usually do little to alleviate economic and social problems (Kedourie, 1960).

Ecological Issues

Ecological problems have become major international issues in recent decades. By the early twentieth century it was evident that growing economic exploitation was inflicting unprecedented injury on the natural environment in the form of increasing denudation of forests, soil erosion and dustbowls resulting from agricultural activity, and industrial pollution. Yet these were problems that national governments could, and in some instances were willing to, control. As technology proliferated throughout the century, increasing use of fossil fuels, the production of electricity using nuclear power, and the problems posed by the growing use of toxic chemicals in industrial processes created novel environmental and health hazards. Efforts to increase food production using chemical fertilizers, insecticides and fungicides, and hormone growth stimulants have become crucial to the success of a capitalist economy, but pose a wide range of demonstrated and suspected health hazards. Chemical products needed to manufacture goods and make machines work spread toxic wastes such as PCBs. Because many of these substances are resistant to decomposition into harmless by-products and spread quickly and widely through the ecosystem, they pose serious problems. What would be the cost of controlling them? To what degree can they be controlled at any cost? And to what degree do they seriously threaten the ecosystem or large parts of it? These issues have become the foci of intense and heated controversy.

Agricultural and industrial threats to the environment are not new. Hunter–gatherers may have played a significant role in the extinction of many animal species (Martin and Klein, 1982). Early civilizations depleted topsoil by overgrazing or salinization that resulted from overuse of irrigation to grow crops. Useful forest covers have long been removed without renewal from hillsides to provide charcoal for metallurgical processes (Hughes, 1975). But never before has so much damage been done simultaneously and in so many ways on the regional as well as the local scale. And never before have humans posed a global threat to the environment. This development would not have occurred without the growth of large-scale economies and more complex technologies, but it would also not have happened had short-term, high profit margins not become the primary concern of business managers. Modern agri-businesses regularly practise monocropping on a large scale, even though this may seriously erode topsoils, in an effort to maximize returns on their investments. They know that if soil exhaustion cuts productivity they can transfer their capital elsewhere and divert depleted land to industrial or residential use, usually making a profit on these deals as well. Hence land that was carefully preserved and nurtured for generations as family farms may be destroyed as the result of a decade or less of industrialized farming (Brown and Wolf, 1984).

There are, of course, developments that might resolve various ecological problems. It is estimated that tropical arboriculture, using fast-growing eucalyptus trees, could, and soon may, supply most of the world's needs for paper while exploiting only a small, densely cultivated land area (Marchak, 1995: 6–10). This would reduce the temptation to cut slowly growing trees in temperate and subarctic portions of the globe and might prove sufficiently economical to drive the traditional sources of wood pulp out of business. Likewise, fish-farming, despite its disease-control problems, is offering a substitute for natural fish stocks that are dwindling rapidly as a result of overexploitation. Like agriculture itself, paper and fish-farming offer effective solutions to dwindling natural supplies, but both require greater investments to produce what is needed. While these approaches have the potential to conserve natural stocks, the domestication of increasing numbers of commodities might tempt entrepreneurs to overexploit natural ocean and forest 'resources' until they become extinct. The development of agriculture and pastoralism has not preserved wild animals but led to their extermination in many regions. Past developments suggest that, in general, aesthetics and a respect for nature take second place to profits and the immediate needs of growing populations. Entrepreneurs are prepared to derive short-term profits from what are judged to be rapidly depleting and threatened resources no less than from ones that are thought to be inexhaustible.

Growing concern with ecological issues has led to the development of conservationist watchdogs and lobby groups, more informed public opinion, and the establishment of government regulatory bodies. Yet governments are generally unwilling to take stringent action or even adequately to consider long-term negative environmental impacts. It is feared that what might be learned would indicate the need for controls that would interfere with short-term economic goals which might enhance a government's chances of remaining in office. Few democratic governments are prepared to think beyond the next election. Despite efforts by ecological lobbies, ecological issues are rarely high enough on the public agenda to have a significant impact on politics. In state-managed economies the desire to promote development for military and political reasons has resulted in economic decisions repeatedly being made without adequate study of environmental impacts or opportunities for objections to be voiced at the local level. The all too predictable result is disasters such as the melt-down of the nuclear generator at Chernobyl in the Ukraine, the drying up of the Aral Sea in former Soviet Central Asia, and the spread of industrial pollution over large areas of the former East Germany.

Technology and Dysfunction

The recent trajectory of sociocultural evolution suggests that future developments will continue to be influenced to a high degree by the course of technological development, especially as this relates to usable energy sources. Continued reliance on fossil fuels may create serious ecological problems as a result of global warming. Yet small-scale, ecologically friendly technologies, such as wind and solar power, seem unlikely to provide adequate sources of power, while atomic energy continues to be both dang-erous and expensive in terms of control and waste disposal. There is also the problem of the depletion of fossil fuels. The more accessible deposits of oil, gas, and coal will in time become exhaus-ted and rising recovery costs will render less accessible ones uneconomical as energy sources. Both the greenhouse effect and the depletion of fossil fuels could lead to the ecological crisis foreseen in the 1960s. This calls into question Gellner's (1988: 224) highly optimistic view that production is likely to go on outstripping population throughout human history.

Despite claims to the contrary, it is no more possible now than it has been in the past to predict the future course of technological development. The creation of cheap fusion power, derived from hydrogen and producing no dangerous by-products, could create enough energy to underwrite unprecedented economic expansion. This would make present fears about our economic future compa-rable to a Palaeolithic hunter's concern that humanity would perish following the demise of the last mammoth. It would also provide the resources needed to begin to cope more effectively with other major ecological problems, such as chemical pollution, acid rain, and desertification.

Yet Richard Adams (1988: 234–42) maintains that, should the energy crisis be surmounted, humanity's problems would not be over. Continuing economic growth would pose increasingly complicated and ultimately insurmountable problems of social integration and control. He questions the idea that, by tapping increasing sources of energy, economies can forever continue to grow at an exponential rate. We cannot predict whether the ultimate curb on progress will be technological or sociopolitical in nature and at what point in human development it might take effect. Yet the accelerating rate of change that has characterized sociocultural development over the past several millennia is not one that can necessarily be extrapolated into the distant future.

If no significantly cleaner and more abundant sources of energy are harnessed, serious problems are likely to arise relatively quickly. Third-world countries that are able to industrialize are unlikely to forgo doing so for ecological reasons. Most people living in developing countries desire a lifestyle that includes material possessions such as television sets and cars. Moreover, transnational industries encourage them to become ever more deeply involved in consumerism as a way to expand world markets. Developed nations seem unlikely to share existing energy resources with developing ones if that means a lower standard of living for their own people. Increasing competition for limited resources is likely to give rise to escalating recriminations and conflicts that will undermine the possibility of effective long-term planning and of the enforcement of conservation measures at the national and international levels. Growing competition will also encourage governments and corporations to resort to novel and often destabilizing behaviour, both within and between countries, in order to achieve short-term economic goals. Disputes over resources may lead to small wars fought with chemical, biological, and nuclear weapons. Sectors of society that feel disfavoured, exploited, or threatened may resort to terrorism to draw attention to their plight, as the Kurds, Muslim Brothers, and some other repressed ethnic, religious, and political groups have long done. Disadvantaged and disoriented individuals may drift in growing numbers toward cults with apocalyptic or nihilistic visions; some of these may resort to violence to punish what they perceive to be an evil and corrupt society, or in an effort to bring about fundamental changes. Even small groups of this sort may cause considerable damage by harnessing ever more powerful military technologies. The combined results of such developments will be escalating social chaos as well as irreparable and cumulative ecological damage.

As the economic order disintegrates, national governments may adopt increasingly irrational programmes in an effort to maintain public support, or find themselves outflanked by political movements that promise salvation by doing so. Examples include the Nazis, who blamed Germany's defeat in the First World War on betrayal by its Jewish citizens and promised to avenge these wrongs, and the Khmer Rouge, who sought to create an egalitarian paradise by slaughtering the rich and the educated. While analogous forms of madness were not unknown in the past, the potential for mobilizing public support and realizing schemes of mass slaughter has been greatly enhanced by modern technology.

While the human mind developed as a means of adapting more flexibly and efficiently to new situations, the ability to imagine countless alternatives and disseminate these ideas through the mass media makes it possible for insanity to become public policy. All too often extreme nationalist movements, such as that of the Nazis, are encouraged by the rich and the powerful in the hope that they will suppress class conflict and protect wealth in societies that are being torn apart by increasing social inequality. Although they depict themselves as champions of democracy, even in developed societies the privileged middle classes have habitually embraced tyrannies and military dictatorships if that appears to be the only effective way to protect private property.

Since the 1930s there has been a growing understanding of the social implications of communications technology. The successive development of printing, mass literacy, and increasingly efficient forms of data processing and telecommunications has altered the ways in which information can be stored and transmitted (Innis, 1951). The development of electronic communications technology has permitted business to be carried on over greater distances and ever more quickly. Each new medium of communication has transformed the image that human beings have of themselves and made possible powerful new forms of propaganda, education, and political organization (McLuhan, 1951). Ever since Gutenberg began printing books in the fifteenth century, political, economic, and religious innovators have been quick to exploit the potential of the written word (McLuhan, 1962). Much of the success of the Nazi Party in Germany in the 1930s has been attributed to the skill with which its leadership exploited the new media of radio and films to promote their ideas. In recent years, growing understanding of the relations among media, knowledge, and social control has permitted the more effective technocratic manipulation of each successive innovation in this sphere, especially by the advertising industry. The illegal circulation of tape recordings of talks by the Ayatollah Khomeini played a key role in subverting the government of the Shah of Iran, and videotapes of Western movies and television serials were likewise important in undermining the authority of the Communist Party in the Soviet Union. Patricia Marchak (1991) has documented how new forms of data processing and communications have been used to promote neo-conservatism and build a new world economy.

We are now witnessing the development of an increasingly complex, multimedia, worldwide communications system. Its expanding networks have the potential for better informing human beings and facilitating discussion, debate, and action on an unprecedented scale. They also have the capacity to be used as instruments of propaganda and for trivializing and crushing dissent. In recent decades the mass media have played a major role in promoting a consumerism that emphasizes short-term, personal gratification at the expense of loyalties to family and community (Grant, 1965; Sanders, 1995). Such attitudes encourage an uncritical participation in an expanding international economy and neutralize a concern with that economy's harmful effects on the individual and on society as a whole.

What happens depends very much on who is able to control the mass media. This makes the outcome of battles currently being waged over national regulation of television broadcasting and control of internet communication of great social and political importance. It may be that in the long run the proliferation of mass media will defeat efforts to use them for purposes of political control. Nevertheless, the effective use of new forms of mass communication to promote violent and destructive political programmes and the increasing tendency for mass media to become subject to monopolistic control pose serious problems. The rapid implementation of neo-conservative economic programmes around the world appears to have been materially assisted by the control of the mass media falling largely, and increasingly, into the hands of companies and individuals who stand to benefit from such a political orientation. The dissemination of views that do not please those in control can be curbed as effectively in a monopolistic free enterprise economy as in a totalitarian state. Hence, even the replacement of a limited range of shared cultural experiences (such as those derived in the 1950s from everyone attending the same weekly movies or watching the same three television channels) by access to a much broader spectrum of individually selected videotapes and over a hundred television channels does not guarantee that people will be well informed about current issues. That depends more upon who produces and controls what the media disseminate.

Another problem with modern mass media, especially the visual media, is their power to blur the distinction between reality and fantasy more than ever before and to shape the images that people have of themselves. Today television viewers can watch endless tales involving violent deaths and murders portrayed in gruesome detail, as well as vivid reconstructions of historical events that often lack authenticity. The media also purvey an image of people as satisfied consumers, not only of material goods but of each other. Originally fantasy was restricted to the spoken word, the written text, and relatively rare works of art. Today motion pictures and television present an ever-increasing number of moving and talking fantasy images. The growing difficulty of distinguishing between fantasy and reality under these conditions may complicate rather than assist people's ability to deal with problems at the individual or group level. While the mass media may help to familiarize people with ways of life and the problems confronting individuals in other parts of the world, there is the danger that what passes for reality is a highly fantasized and distorted account. The increasing realism with which those who control the mass media can objectify fantasy poses serious problems for socializing children and conducting productive political debates (Solway, 1997).

An equally serious problem is how to deal with the ever-increasing amounts of information that, as a result of proliferating communications technologies, are saturating and threatening to overwhelm people's lives. Some of these messages are generated and diffused independently by individuals and small groups, the rest are filtered through media controlled by governments and transnational companies. All of them are permeated with biases and agendas which may be, but usually are not, made explicit. Some of them seek to be truthful, some do not; some are valuable and even essential, most are irrelevant and worthless. All of them seek to attract our attention and influence our behaviour.

The greatest challenge we all face is learning how to discriminate between information that is relevant to our needs and information that is not. Unless people can do this effectively, their judgement and critical faculties are in greater danger than ever before of being overwhelmed by an avalanche of messages. The key to successful lives, careers, and citizenship in the hectic modern world is our ability to transform information into understanding. All of us face the daunting challenge of learning how to discriminate among increasingly superabundant information, how to evaluate novel proposals ever more swiftly and effectively, and how to distinguish between what is valuable to us as individuals and citizens and what is harmful. At present, as the criminal exploitation of historical animosities to promote political factions in the former Yugoslavia and in Central Africa clearly demonstrates, the problems appear to be growing faster than do solutions. Such challenges strain the abilities, evolved by our hominid ancestors, to deal rationally with our social as well as our natural environment. Part of the answer must lie in learning to find time for reflection and analysis. Only in this way can data be transformed into the personal understanding that is essential for coping successfully with change.

It is increasingly accepted that all human groups share a common, biologically grounded human nature (Carrithers, 1992). It is becoming possible, however, for mind-controlling drugs and genetic engineering to create human beings who are physically or psychologically adapted to perform specific tasks or play particular roles in society. Long ago slave owners and philosophers such as Aristotle self-interestedly fantasized that some individuals or groups were biologically destined to be slaves, while others were naturally suited to be masters. Historical and ethnographic evidence proves that no human groups either want to be slaves or are slaves willingly, as the draconian measures needed to control such victims of oppression and exploitation clearly demonstrate. Today, however, the possibility exists for the first time of being able to create individuals biologically programmed to risk their lives or perform repetitive tasks that no one could be expected to do except as the result of intensive indoctrination, coercion, or desperate financial need. Creating people programmed to be slaves, soldiers, or uncomplaining factory workers would release sociocultural evolution from the constraints of a common human nature forged during the pre-agricultural past, when higher primate biology was altered to enhance hominid ability to manipulate symbols and perform an expanding range of tasks. This also involved considerable modification of the biological basis of sexual behaviour to make sexual pair-bonding the basis of a large number of activities, such as child care and food processing. The slow rate of cultural change at this period facilitated natural selection, which adapted hominids for living in groups of a few dozen to a few hundred individuals who subsisted by means of hunting and gathering (Carrithers, 1992; Woolfson, 1982).

The ability to alter human behaviour by means of drugs or genetic engineering would make it far easier for those who control such technologies to treat other human beings as a means for achieving specific ends than has been the case so far, when only political manipulation could be used to shape human behaviour. If applied on a sufficiently large scale, it could deliver the future of humanity into the hands of a small number of people who would manipulate others for their own selfish and short-term goals. Making it possible for a small number of people to impose their wishes and objectives on all human beings would place the future of humanity in much greater jeopardy than it has ever been in the past.

All cultures have perceived good and bad elements in human nature, although there has been no agreement about how various forms of behaviour should be classified or which ones predominate. Efforts have been made to curb undesirable behaviour and to reform human conduct by employing a mixture of example, moral instruction, persuasion, threats, and restraint. But, until now, such efforts have been limited by our inability to alter human biology to any significant degree. Because of that, what human beings *are* biologically serves to limit cultural diversity. While human behaviour has been selected to emphasize learning and openness, its malleability has been kept within bounds by a balance of tendencies that promote, among other features, both co-operation and competitiveness. As a result of technological advances, a biological bulwark that throughout history has protected human beings from some of the extreme dangers posed by the flexibility of their learned patterns of behaviour is in danger of being breached.

What is to be done? Should further scientific research relating to the modification of human behaviour be forbidden? Since the First World War the military use of chemical and biological warfare has been banned and largely curtailed, but the development of more sophisticated weapons of these sorts has not been halted. There is little to suggest that societies can prevent the overt or clandestine development of new technologies if it is in the interests of powerful individuals, companies, or governments to do so. Totalitarian regimes might find it especially attractive to create people predisposed to accept and carry out their policies. Yet, from the 1890s into the 1940s, democratic governments in North America and elsewhere were prepared to intervene biologically by involuntarily sterilizing young women who were judged to be of inferior intelligence or to have behavioural problems (McLaren, 1986). Today similar democratic governments, faced with growing communal disorder as a result of their neo-conservative policies, might be tempted to stigmatize politically troublesome elements as psychotic and take drastic action to 'cure' or 'neutralize' them. In the future, if increasingly sophisticated forms of propaganda prove inadequate to muster support for their policies, even nominally democratic governments might resort to sophisticated forms of genetic engineering or *Brave New World*-style therapies on an ever-expanding scale in an effort to adapt people to public policies rather than formulating public policies to suit human needs. Such action could be justified on the grounds that it promoted human happiness, curbed violence, or was necessary to prevent being overtaken by societies that had already adopted such practices.

Finally, attempts to curb scientific research will be met with earnest claims that science has the right to investigate any topic. Research on genetic engineering will be justified on the grounds that it can provide techniques to eliminate genetically based ill-nesses, enhance intelligence, promote better mental health, and make individuals physically more attractive. All will be offered the opportunity, if they can afford to pay for it, of being the kind of person and having the kind of children they want. The dangers inherent in the application of these technologies are only likely to be explored, and ethical systems devised to try to control them, after their development is far advanced. By that time, given the complexity of the world economy and the limited powers of jurisdictions to curb human behaviour, it might be too late to stop either the further development of dangerous technologies or their large-scale abuse. This is the sort of problem that Hornell Hart (1959) covered with his law of cultural lag.

Guidance from the Longer Term

While noting that what is going on at present differs in many important respects from what happened in the more remote past, it remains to be considered what may be relevant for understanding the future from longer-term sociocultural evolutionary trends. All that is known about physical and biological evolution indicates that studying the development of the natural and biological universe, which from a scientific point of view appears to have been a purely fortuitous process, cannot reveal anything about human destiny.

Humans are unable to look beyond themselves for guidance. Only their culture and their individual consciences (often shaped by deeply held religious beliefs) can be a source of moral values (Bowler, 1984: 316). Yet in the long run sociocultural evolution has been shaped by the more successful and life-enhancing choices that human beings have made throughout history. Hence, when duly considered, such beliefs may be of value for directing future conduct. In addition, past growth patterns often control the appearance of new characteristics and hence provide some direction to subsequent trends.

It is still debated whether technological change occurs continuously or is episodic and opportunistic. Few social scientists would now accept the Enlightenment idea that humans constantly invent more elaborate technologies in an effort to control nature more effectively and make their lives richer and more secure. Technological change is viewed as a response to ecological pressure and to various forms of competition within and between societies. It is also recognized that innovation, although guided by rational foresight, does not occur in a linear fashion but produces many dead ends and has happened more quickly and had greater social impact as technological knowledge and human populations increased.

Yet the factors that bring about technological change are sufficiently uniform and the social transformations that result from them sufficiently influential, especially in the context of intersocietal competition, that they create a general evolutionary pattern. Linear trends in technological and social evolution are apparent in the similar sorts of societies that developed in roughly the same order in different parts of the world, even in quite different ecological settings. A general similarity has long been observed in cultural trends in the Old World and the Americas, despite a general dearth of potentially domesticable animals in the western hemisphere. Yet early civilizations first appeared in river valleys in arid regions, mountain valleys, tropical forests, and humid temperate areas. These similarities appear to result from limitations in the paths that technological development can follow, in the viable forms of society that any particular level of technology can support, and in the sorts of societies that can survive in competition with alternative forms (Trigger, 1993).

Another general feature of sociocultural evolution is the expanding scale and diversity of human interaction. Over time larger and more complexly integrated societies develop. These are able to affect the ecosystem more dramatically, influence much larger surrounding areas economically and culturally, and impose their will on smaller, less tightly integrated societies. The increasing capacity of technologically advanced societies to shape the development of neighbouring, smaller-scale ones, and often to absorb them completely, helps to impose a linear direction on sociocultural development. Because of the increasing rates of technological and social change associated with more complex societies, there appears to be less opportunity for significantly different types to exist alongside each other for long periods when a complex society is involved than when all the societies are small-scale ones.

The larger scale and greater resources of technologically more advanced societies permit them to cope easily with short-term natural disasters that might physically annihilate hunter–gatherer bands. So far, however, this protection has been assured only at the expense of harder work, greater discipline, more elaborate social hierarchies, and significant economic inequality. Moreover, large systems suffer from their own problems. Often they inflict vast and escalating damage on the environment. While the detrimental ecological impacts of human activities, as we have already noted, were once local and regional in scope, they are now destabilizing at a global level. The collapse of complex societies as the result of failure to cope with ecological, administrative, or political crises can inflict great suffering on vast numbers of people (Yoffee and Cowgill, 1988).

But, unlike China, where dynastic crises have produced massive political and economic dislocations every few centuries, there has not been a major organizational collapse in Central and Western Europe since the end of the Roman Empire. This stability appears to have been largely a beneficial result of Europe's political decentralization. A collapse nearly occurred in Central Europe during the seventeenth century as a consequence of the Thirty Years War. A more extensive collapse might have happened during the first half of the twentieth century as a result of the misuse of increasingly destructive military technologies in the course of two world wars. Moreover, even the repeated political crises that have disrupted China, the Middle East, and India over the centuries have not prevented the continuity of the civilizations of those regions, although periodically their prosperity has been under-mined and much life lost. The losses do not appear to be nearly as complete as those that accompanied the collapse of the earlier Indus Valley or Hittite civilizations. Keeping complex societies economically and socially integrated and responsive to environ-mental challenges requires a flexible social and political organization. The more effectively a society can monitor its situa-tion and translate that information into appropriate action, the better able it is to cope with a broad range of challenges.

Planning or Laissez-faire

There has been a lengthy debate in Western society concerning the effective management of technological and social problems. One side has argued that no overall planning is best, and insisted on interpreting efforts to regulate society as unwarranted interference, if not outright tyranny, as well as a source of inefficiency and wastefulness. This side maintains that, because of the limited knowledge of a society that even the most informed planners possess and also because of their self-interest, social planning is incapable of providing for the best interests either of society as a whole or of its individual members. They believe that internal diversity, individual preferences, and individual decision-making offer the broadest basis on which the social equivalent of natural selection can operate and the widest range of personal tastes and preferences can be satisfied. An 'invisible hand', rather than the collective exercise of human intellect, will ensure the long-term well-being of society. This argument is used to justify a *laissez-faire* attitude towards economic and social development and to argue that in the long run a consumer society is more capable of surviving, prospering, and bringing about human fulfilment than is a planned one.

Yet, with equal conviction, other thinkers have seen rational planning as a highly effective way to serve the best interests of society. Although most Enlightenment thinkers, especially in Britain, supported *laissez-faire* policies, the Enlightenment's rationalistic view of human behaviour can be construed as supportive of planning. If human beings share an essentially similar nature and therefore have a relatively narrow range of preferences, as rationalists believe, this should make planning more efficacious than if tastes and values are highly varied even within the same culture, as romantics and *laissez-faire* philosophers assume. The collapse of the Soviet Union has been interpreted as offering empirical proof that centralized planning does not work. This claim, however, overlooks the possibility that its collapse may not have been caused by planning as such, but by widespread corruption and a lack of adequate planning and administrative skills (Dunn, 1993: 124).

In Western society there are also radical disagreements about what constitutes cultural diversity and choice. These views cross-cut the political right and left and combine individualistic and communitarian options in ways that suggest that traditional ideologies are indeed becoming fragmented. Within the context of the consumer society there is widespread encouragement to equate diversity with the number of brands of toothpaste or toilet paper on sale in supermarkets, or the number of candidates competing for public office, whether or not these have anything different to offer. More profoundly, however, for many liberals diversity and choice imply an enhanced capacity to define and realize a personal lifestyle and to obtain employment that is congenial and fulfilling. Such a potential for self-realization is associated with societies that have organic unity, as Emile Durkheim (1893) defined this concept. In these societies, people are linked by mutual dependence resulting from economic specialization. Hence individuals are freed from the need to share a complex belief system in order for a society to hold together.

On the other hand, many supporters of cultural diversity and collective rights regard the kind of diversity found within consumer societies as a sham. It is, moreover, a sham that erodes established local and regional cultural differences that might impede the spread of a mass industrial culture and thus prevent businesses from increasing profits by exploiting as large a market as possible. The conservative philosopher George Grant (1965) argued that the expansion of a consumer society relies on advertising to break down regional barriers, eliminate local competition, and promote the standardized products of transnational corporations. This involves the elimination, as far as possible, of distinctive local, regional, and national cultures and with them whatever wisdom has been achieved as a result of smaller collectivities adapting to local conditions over long periods of time. Even radical cultural expressions of social protest, such as punk music, are quickly co-opted by entrepreneurs. This results in such movements being trivialized and rendered harmless as expressions of dissent as well as becoming a source of profit for transnationals.

Grahame Clark (1986) argued that, under a façade of enhanced diversity and personal choice, modern industrial societies are imposing a stultifying uniformity of thought on people, to an ever-increasing degree. Anthropologists, such as Lévi-Strauss (1985), likewise express grave concern about the way in which industrial societies are eroding long-established regional diversity and destroying band and tribal societies around the world. This is seen as erasing knowledge that is potentially of great value for the future well-being and progress of humanity, in the same manner as the spread of patented seed crops and the cutting down of tropical forests for beef production are destroying age-old reserves of plants that also might be of great value to humanity.

Yet, while the creation of a world economy is eroding the linguistic, ethnic, and religious blocs that over the course of human history have formed diverse cultural mosaics in every region of the world, it is also promoting the development of cultural pluralism and multiethnic states. As a result of massive population move-ments, each of these multiethnic states is coming to contain within itself much of the cultural diversity once found scattered in isolated pockets in every part of the world. While religious, ethnic, and linguistic affiliations continue to provide individuals with a sense of identity within these larger economic and political units, the significance of this diversity is radically transformed and rendered politically less important. The role of the nation state as the expression of the collective concerns of a distinct people is rapidly declining. The creation of larger political and economic entities is limiting and usurping the traditional powers of the nation state. At the same time, growing multiculturalism makes it possible for people from many different backgrounds to live alongside each other and use their distinctive traditions to work together more effectively for their personal and collective benefit. Such multi-ethnic societies, because of their more widely ranging connections and greater total knowledge of different cultures, have a competitive advantage over traditional nation states in carrying on business and hence prospering in an ever more cosmopolitan world. Ethnicity is ceasing to be equated with the nation state.

Knowledge as Power

Another major feature of cultural evolution is the development of an ever more detailed corpus of knowledge which necessarily has to be shared by a growing number of specialists. There is also a tendency for this knowledge to be evaluated in an increasingly instrumental fashion in terms of its ability to achieve specific goals. Romantics lament this development as the loss of a holistic perspective and as a source of spiritual alienation. Rationalists tend to value an increase in practical knowledge because they believe that it results in the more objective understanding of the natural world and of human behaviour, and therefore in a more profound and accurate comprehension of the possible consequences of human action.

As a consequence of the evolution of knowledge, a supernatural understanding of causality has slowly given way to more natural understandings that allow more sustained and continuous economic and social development. Jürgen Habermas (1979, 1984) has argued that societies evolve through a strict sequence of world views, each of which is able to produce more adequate cognitive and moral structures than was its predecessor. Over time sensory data are recognized as the principal, or even the only, source of knowledge concerning what is real in the natural world. Norbert Elias (1978: 254–60) maintains that a central feature of modern, as opposed to earlier, forms of thought is the ability to conceive of natural processes as constituting an autonomous sphere acting in a purely mechanical fashion without intentions, purposes, or destiny, and as having meaning for humans only if humans have adequate knowledge to control such processes and hence assign their own meanings and purposes to them. The demise of a belief in an anthropocentric universe also promotes the realization that nature is indifferent to human beings and must be dealt with technologically. According to Julian Huxley (1960: 54), gods, like earlier concepts involving the animistic projection of human mental and spiritual qualities into non-human nature, are ceasing to have interpretative value.

Huxley (1960: 242–3) also argues that the development of a naturalistic understanding of the universe involves replacing traditional dualistic modes of thought by unitary ones. Hotness and coldness are recognized as states of temperature and lightness and heaviness as states of weight or mass. Values are no longer seen as absolute and God-given but are recognized as arising from experiencing the kind of behaviour that promotes the well-being of particular societies and their members (Gellner, 1988: 190–1). It is likewise recognized that there is no permanent ontology. Instead, our understanding of reality changes as a result of our growing knowledge of how to manipulate nature. Eventually it is realized that our reality consists of what we know, rather than what actually is, and hence that reality is forever revisable. Gellner (1988: 204) believed the transition from an anthropocentric to a modern form of cognition to be irreversible; once a modern understanding had been achieved, the way back was forever blocked. While this view does not take account of current movements, such as New Age philosophy, which seek to revive a mystical, anthropocentric view of nature, Gellner was betting that in a modern technological society such beliefs could never again become dominant. Susceptibility to astrology and other forms of magic and divination demonstrates, however, that anthropocentric views of nature continue to exert considerable attractions, and one wonders under what conditions of severe social stress such views might again predominate. Gellner clearly was not thinking of such pathological societies.

The opposite danger is that recognition of humanity's animal origins, combined with a growing temptation to treat individual humans as yet more objects suitable for technological manipulation, might lead to a loss of proper respect for human beings and a tendency to ignore or undervalue the importance of the subjective qualities of the human mind. The idea that technology can provide quick fixes for all personal and social problems, the loss of faith in all human beings sharing common characteristics and rights that resulted from the inroads of racism prior to 1945, and a growing sense of individual weakness when confronted by ever-expanding economies and interventionist political systems have helped to erode a sense of human worth. This has encouraged the belief that, because human beings are part of the natural realm, they are no longer worthy of respect; nor should they be more immune from exploitation than any other resource. The idea that human beings are not set apart from the rest of nature in any fundamental way helped to create a climate of thought in which it became possible for Nazi doctors to use living people for their lethal medical experiments, just as earlier it had encouraged various forms of egocentric, antisocial philosophy, such as that of Friedrich Nietzsche. Similar instrumental views of human beings are evident to a lesser degree in most behaviourist approaches to human psychology and social analysis. Because this misuse of positivism ignores the special reflexive qualities of the human mind, it gives rise to bad (or at least incomplete) science as well as to dangerous philosophy (Bunge, 1996).

Expanding Social Responsibility

Evolutionists have assumed that in the past local groups that resisted change, or were not able to develop as quickly as others, were unable to halt the formation of greater sociocultural complexity elsewhere. Failure to develop condemned such groups to relative powerlessness when they had to compete for resources with more complex societies. Recently some relativists have made a concerted effort to deny this. They have drawn attention to the spirited resistance that North American Indians made against European settlement and attributed European success almost entirely to the differential lethal impact of epidemics of European diseases to which aboriginal Americans had no acquired immunity (Dobyns, 1983; Merrell, 1989). It is true that the role of these diseases was greatly underrated in the past. But an aboriginal desire for European goods, as well as the Indians' not having evolved political structures that were large and tightly integrated enough to cope with European encroachment, also played a major role. From the early seventeenth century on, the Indians recognized that their small political units and internal factionalism made it difficult for them to resist European settlement. Yet it was not until the 1820s that a single Indian nation had developed the concept of a national territory and was prepared to impose the death penalty on individual citizens who alienated land without the approval of their government (Green, 1996: 515). Even this political development did not save the Creek Indians from having their national territory seized by the United States less than a decade later and they themselves from being deported west of the Mississippi River. However tenaciously less advanced peoples have resisted domination by more numerous and technologically more advanced neighbours, they have had either to become more like their neighbours or to succumb to varying degrees of domination and exploitation by them. Although in the earliest phases of colonization Europeans frequently had to treat Indians as political equals or even defer to their wishes, successful colonization quickly freed them from such ties of dependence and allowed them to reduce native people to powerless enclaves.

Another controversy relates to whether the link between material wealth on the one hand and power and high status on the other is socially constructed or innate in human nature. If human beings are naturally acquisitive and power-seeking, the desire to outdo others would have to be universal, even in hunter-gatherer societies. As noted in chapter 9, in small-scale societies both the sharing of resources and egalitarian behaviour are reinforced by positive sanctions, such as praise and respect, and by negative ones, such as joking, gossip, ostracism, and accusations of witchcraft. All these mechanisms rely on public opinion for their effectiveness. The fact that coercion is required to enforce sharing and self-effacement in these societies indicates that these qualities do not prevail spontaneously in human behaviour. Primate behaviour also suggests that efforts at self-promotion characterized higher primates, and especially male higher primates, long before the development of symbolically mediated behaviour in the hominid line (Wrangham and Peterson, 1996).

In larger societies public opinion no longer suffices to maintain sharing of resources and curb self-assertion. One of the main duties of the state is to protect private property and punish subversion. The dominant classes are in an excellent position to control the dissemination of information in such societies and hence to curb dissent. Prosperous individuals no longer have to redistribute their wealth in order to avoid accusations of witchcraft or falling victim to the witchcraft of others as a result of their lack of generosity. Instead they threaten lower-class people who are not sufficiently deferential with accusations of witchcraft and execute those whom they imagine may be trying to harm them supernaturally (Trigger, 1990).

This raises the question whether any conceivable method can constrain greed in the larger societies of the future as was done in the smallest human societies of the past. While we should not minimize the role played by idealism in the creation of social democratic societies earlier in the twentieth century, the welfare networks that were constructed or greatly expanded throughout the Western world after the Second World War were built on habits of sharing limited resources that were established during that war and were later sustained by fear of communism. This did not lead to economic equality but did create a safety net that to varying degrees promoted a more even division of wealth in Western societies during a period of considerable prosperity. Less generous attitudes prevailed, especially among the wealthy and powerful, as economic expansion slowed in Western Europe and North America in the 1970s (Krieger, 1986). No society that either from necessity or by popular choice opts to promote greater economic equality can assume that this policy reflects human nature. Equality needs to be defended by public opinion and public sanctions no less than property and individual safety need to be protected by the power of the state. It is not clear how or to what extent this can be done in large societies.

The development of capitalism has promoted the emergence of societies composed of open classes rather than endogamous orders. More social mobility is required to accommodate a rapidly changing and expanding division of labour. A more complex division of labour also necessitates more geographical mobility, which weakens traditional structures, such as communities and extended families, and emphasizes the importance and self-reliance of the individual. The aims of widespread public education, which accompanied industrialization, were to encourage literacy and the development of other basic job skills and promote a broader sense of community in the form of nationalism (Gellner, 1983). Nationalism was built around a core of old ethnic loyalties or simply being the citizen of a particular state. It was not the only way to unite populations characterized by enhanced individualism, but it was very effective for diffusing internal conflicts, especially between classes, by claiming that internal social problems resulted from foreign competition. Nation states have also been the only entities able even to a limited degree to curb the predatory behaviour of powerful businesses in the interest of society as a whole.

It is generally assumed that individuals are most loyal to groups that are physically closest to them, such as families, communities, regions, linguistic groups, and nations. The question remains: how far can such loyalties be stretched? Does a broadening of loyalties inevitably result in a diminution of allegiance to local and regional entities? So far there has been only limited development of transnational loyalties, often among people who share the same religion or to a lesser degree do business with one another on a regular basis.

Yet, as the number of skilled occupations increases, each develops its own traditions, rules, associations, and understandings. Members of different professions remain loyal to bonds of class, religion, politics, ethnicity, region, nation, and culture. Yet it is probably true that Russian and American physicists, chemists, and medical doctors have at least as much in common with each other as they have with plumbers, bricklayers, or lawyers in their own communities. They co-operate in building international associations, in exchanging scientific information when not prevented from doing so by their governments, and sometimes in addressing major international issues, as Physicians for Social Responsibility have done in promoting nuclear disarmament (Aronow, 1963). They have also demonstrated the ability to co-operate in providing aid following major disasters, such as the Chernobyl nuclear meltdown. This kind of internationalism has long been inherent in craft knowledge and scholarship (Childe, 1958), but is now growing as a result of rapid increases in knowledge and faster means of travel and data exchange.

Elias (1982: 88) has argued that, as societies move to levels of integration of a new order of magnitude, they do so in conjunction with more differentiated social functions. As a result, not only behaviour but the totality of emotional life and personality structure changes. As the social fabric grows more intricate, individual self-control also has to become more differentiated, better rounded, and more stable. According to Elias, little compelled medieval people, perhaps least of all members of the upper class, to impose constraints on their behaviour. As a result they frequently conducted themselves in a wild and even cruel fashion (1982: 72). Modern people, by contrast, require far more self-control to manage their technology and complex social relations. Elias maintains that these skills began to be developed as feudal warriors were transformed into courtiers.

The relation between more complex societies and greater self-control is a functional one. As societies increase in scale and public opinion ceases to be as effective and general a form of social control as it was in smaller-scale societies, increasing self-control becomes essential to cope with more complex technologies and their social consequences. This does not guarantee that such forms of self-control will develop, but without them any overall increase in social complexity remains problematical. Thus the randomness of sociocultural evolution is not curtailed by some mysterious orthogenic force, but by the inability of systems to function at increasingly complex levels of development without evolving some specific regulatory mechanisms. Far from stating the obvious or implying that whatever exists in the sociocultural realm does so because it works or is the product of some form of ecological determinism (Smith, 1973), functional explanations account for much of the significant cross-cultural regularity found in societies at similar levels of complexity.

Elias (1982: 241, 271) also stresses that individuals had to acquire greater foresight as they came to depend on ever larger numbers of people. That required continuous reflection, more precise and articulate regulation of one's own feelings, and greater knowledge of the social and natural terrain. Fluctuations in behaviour and feelings did not disappear but became more moderated and controlled. Generally speaking, modern everyday life is less subject to radical changes of fortune, and physical violence tends to become remote from the lives of most people (1982: 238). Elias notes, however, that rationality and inhibition of drives can debilitate as well as assist people. Tensions once resolved, often violently, between individuals or groups must now largely be dealt with internally. As social life becomes less of a danger zone, the self becomes more so. These tensions create dissatisfaction, restlessness and boredom, as well as promoting idiosyncratic behaviour (1982: 242, 298).

While Elias's ideas may seem naive at the end of a century marked by extreme violence and the disruption and destruction of human lives on an unparalleled scale, he does describe the nature of life for an ever-increasing number of people in the more developed societies. Even in authoritarian societies that have advanced technologies, the specialized knowledge possessed by growing numbers of individuals greatly increases the social cost of killing them on account of their political or religious views or their ethnic identity. That does not rule out, however, the violent suppression of dissent, especially in societies destabilized by economic decline. In Chile, Argentina, Indonesia, and many other countries, this repression has been directed against the political left. Major repression has also characterized some societies following a policy of planned modernization. Soviet officials tried to limit the economic losses from such repression by putting skilled dissidents to work in prison camps or Siberian exile, doing the same jobs they had prior to arrest. That does not, of course, explain the large numbers of dissidents who were executed.

The wasting of human talents as a result of political oppression is only one of the problems encountered by repressive societies when they have to compete with more open ones. Also crippling to the Soviet economy was the ideologically motivated support for Trofim Lysenko's Larmarckian approach to biology, which severely retarded the development of domesticated plants better suited to the climate of Siberia and Soviet Central Asia, and thus held back Soviet economic development. Problems of coercive control are also evident in retrospect in the extent to which the Nazi planning of the German war effort turned out to be less effective than was the less centralized direction of the Western wartime economies. In societies that depend on an increasingly diversified body of knowledge to function, respect for individual contributions is essential for promoting progress.

It is often suggested that the current situation, in which about one-fifth of the world's population is relatively affluent while the rest remains desperately poor, is a passing phase that will be remedied by worldwide economic growth. Neo-conservatives argue that the trickle-down effect will raise the living standards of even the poorest people in societies that have expanding economies. History and technological development do not provide grounds for such optimism. Although overall productivity has increased greatly since the start of the industrial revolution, wealth differences between the north and south have increased. It is suggested that the rapidly escalating cost of the technological infrastructure required for full participation in the new electronic global economy will continue to widen the gap between rich and poor countries (Elliott, 1996). This is unlikely to be offset by the anticipated emergence of a worldwide, internationalized middle class as a result of the spread of a global economy. Economic development has resulted in an earlier gap in wealth between a tiny ruling elite and a vast number of poor people being replaced by a more open hierarchy in which more people at the top share wealth. But the overall distribution of wealth varies from one society to another depending on the economic development of various regions and on the mechanisms that the political parties in power use to tax and redistribute wealth. In the West, as the mechanisms for redistribution have been dismantled by neo-conservatives, the distribution of wealth has become noticeably less equitable than it was in the 1960s.

Conclusion

Neither the general panorama of sociocultural evolution nor a more detailed perspective on European history over the last few hundred years, or the last few decades, provides a clear view of the future. This is because social change is complex and the future has not yet happened. Yet an understanding of what occurred in the past provides useful insights into some of the basic problems we are facing and may help to distinguish potentially fruitful efforts to resolve these problems from self-defeating ones. Having developed a technology that has freed it from direct domination by natural forces, humanity today finds itself more menaced by the techno-logical environment that humans have created for themselves than it ever was by the natural environment. The destructive impacts of technology on the environment are only now beginning to be understood, and the constantly accelerating speed of technological change is creating social, cultural, and psychological problems of a new order. The increasing tempo of technological change also hastens social and cultural change, rendering time-tested and relatively benevolent traditional knowledge less efficacious as a form of social control. That encourages experimentation to dis-cover new ways of doing things, which in turn gives greater scope to individual rapacity and to runaway copying of role models who appear successful in the short term but may not be so over a longer period. All of this creates dangers and dislocations at the individual, national, and global levels. Yet population pressure alone is sufficient to preclude turning back to some simpler style of life. That could only come about as the result of a tragedy that would produce a demographic decline of unimaginable proportions. Such a catastrophe would leave the survivors in a world that might be ecologically ill-equipped to support them and without the knowledge they would need to cope with their new circumstances. Given the unacceptability of that alternative, the fundamental challenge that confronts all human beings is to learn how to subject an all-encompassing technology to human regulation.

11
Evolution and the Future

Today evolutionary studies confront an extraordinary irony. Social change is occurring faster than ever before and increasingly powerful technologies are uniting the various trajectories that have characterized human development in different parts of the world to create a single, interconnected, though still multistranded, process of change. All of this is occurring under the aegis of a dominant neo-conservative ideology. Simultaneously, evolution is being denounced as a fantasy that was invented to justify colonialism, social injustice, economic exploitation, slavery, gender oppression, cultural elitism, and almost any other abuse that can be imagined (Hodder, 1991; Shanks and Tilley, 1987; Terrell, 1988; Tilley, 1995). It is generally acknowledged that sociocultural change has not been continuous or without setbacks and that it has not resulted in improvements in all aspects of human life or benefited everyone equally. In recent years few evolutionists have claimed that it has. Evolutionists have also recognized that such directionality as is evident in sociocultural change is not the unfolding of a cosmic plan, as many eighteenth- and nineteenth-century thinkers believed, but the outcome of the relative capacity of societies organized on different principles to influence each other. Is sociocultural evolution a self-serving fantasy or does it refer to something real? How people answer that question relates closely to how they attempt to deal with the future.

All social scientists recognize that consciousness plays a role in bringing about sociocultural change, although they see it operating in different ways. Human beings are able to fantasize, conceptualize alternative courses of action, criticize their own ideas and those of others, and debate with each other. Their capacity to devise complex strategies of purposeful action endows their behaviour with a unique degree of intentionality and continuity.

This adds a genuinely teleological dimension to human behaviour that differentiates it from the way the rest of the cosmos and the animal kingdom operate. While it would be going too far to claim that the whole of human history is a project of intentionality, consciousness endows at least short-term sociocultural change with purpose and direction. To try to encompass the study of human behaviour within the methodology of the natural sciences, by viewing it as if it were controlled entirely by forces that lie beyond humanity's conscious control, would be to ignore the unique properties of human behaviour that have emerged as a consequence of humanity's exceptional powers of self-reflexivity. Yet there may also be a temptation to ascribe too prominent a role to the human imagination.

Seeking to Abolish the 'Tyranny' of Science

Postmodernists seek to negate the power of science to control human life by arguing that all discussions of human behaviour are hopelessly contaminated by subjectivity. They maintain that what is presented as knowledge of human behaviour or human history is designed to serve the interests of its creators. Such knowledge promotes partisan causes by making their goals appear natural and inevitable, misrepresents self-interested behaviour as altruism, and helps powerful individuals convince themselves that their selfish conduct is in the public interest (Barnes, 1974, 1977; Feyerabend, 1975; Habermas, 1971; Marcuse, 1964). In recent decades, it has become popular to argue that sociocultural evolution is a set of myths that has been invented to devalue technologically less-developed cultures and suggest that the only constructive role these cultures have to play in the modern world is as examples of how more advanced peoples might have lived in the remote past. These claims, it is argued, are in turn used to justify colonialism, the seizure of aboriginal territories, efforts to 'civilize' native peoples against their will, ethnocide, and even genocide. It is alleged that, while sociocultural evolutionism was devised in the eighteenth century as an ideology that would empower the efforts of the Western European middle class to free itself from the constraints of political absolutism, it evolved, with the growing power of the middle class, into a justification for Western hegemony and oppression. There is currently a romantic tendency to idealize small-scale, low-technology societies as being affluent, peaceful, and ecological; and static rather than rapidly changing societies as well adapted to the natural environment and in tune with unspoiled human nature. Such ideas resonate equally with New Age theology and a literal reading of the ancient Judaeo-Christian myth of the Garden of Eden.

Much of this relativist critique of sociocultural evolution is insightful and valid. Evolutionary studies have been associated both consciously and unconsciously with varied social and political agendas and have become permeated with ethnocentric prejudices. Their implication that change must inevitably lead human beings onward and upward does favour change over stasis and denies the possibility that a commitment to stability may be an equally significant achievement and one worthy of respect in its own right (Mintz, 1985: 145). Applying the concept of progress to whole societies further implies that more complex cultures are superior to less complex ones. Ever since the eighteenth century, critics have pointed out that evolutionists tend to overlook the fact that technological development frequently produces social and political changes that are prejudicial to the well-being of large numbers of people.

All these objections, however, refer specifically to the political uses that are made of evolutionary concepts. They do not address whether or not there is a direction or shape to human history or how such a shape might be explained. In a paradox that confronts all relativist critiques, anti-evolutionism has ideological implications that are no less striking than the ones it ascribes to evolutionism; hence, judged according to its own principles, it, no less than the theories it condemns, can be subscribed to only at a moral cost. Finally, in their honest desire to defend the integrity of older and smaller cultures, extreme anti-evolutionists have to deny the possibility of genuine progress and therefore of overcoming deep-seated problems and creating a better type of society than has ever existed before. The flip side of respect for all cultures is resignation about the possibility of things ever being done better. Fortunately, these two approaches are not irreconcilable, since evolutionism and relativism can be construed as complementary, rather than antithetical, concepts.

Many moderate scholars argue that, while the problems chosen for study and the way data are interpreted are inevitably influenced by subjective factors and hence by the social environment, by paying careful attention to the evidence scholars can achieve a more objective and comprehensive understanding of the past (Bunge, 1996; Trigger, 1989; Wylie, 1993). Public awareness of such findings has a positive role to play in making choices that at the individual and collective level are vital for the future of humanity.

Mythistories

The historian William McNeill (1986) has argued that, as Western civilization has abandoned fixed religious and philosophical concepts in the context of ever more rapid social change, these concepts have been replaced by flexible and changing formulations that seek, on the basis of what is known at any period about nature and human behaviour, to make sense of the human condition as guides for social action. An example of such a formulation would be the tenets of the Enlightenment. McNeill calls these provisional formulations 'mythistory'.

As the term indicates, McNeill does not claim for such formulations the status of absolute truths. But he does maintain that they must accord with the best current knowledge if they are to be useful. Yet, of necessity, actions must be based on values as well as on what are held to be facts. Furthermore, since knowledge is distorted by bias and constrained by the limitations of human experience, it is never wholly rational or accurate. At their best, specific formulations provide guidance appropriate for particular levels of technology and social organization. To avoid the worst shortcomings, mythistory must aim to be general in its application, since it can properly serve the long-term interests of specific groups only by seeking to promote the welfare of all humans. It is also by definition relative, since beliefs that constitute a useful guide at one stage of development tend to become increasingly dysfunctional, and even maladaptive, as societies change. Myths that are constructed on erroneous or antiquated beliefs, or self-centred values, may prove disastrous for those who act on them.

According to McNeill the only objective test of mythistory is its ability to promote the welfare of believers. As Childe noted, the world that societies adjust to is not the world as it is but the world as people imagine it to be. Yet he maintained that the two must correspond to a significant degree if societies are to survive. It therefore seems likely that beliefs based on the most accurate and advanced scientific knowledge will be the most efficacious, since they correspond most closely to the realities to which human groups must adjust. While mythistory tends to be formulated by intellectual elites, it must be accepted as a basis for action by a substantial cross-section of a population. It is therefore a mass phenomenon. McNeill discusses only hegemonous mythistories, but in complex societies there are usually countermyths competing with the dominant one.

Thus it appears that both the creation and application of knowledge are influenced by myths. What is believed to be true influences the questions that scientists as well as other people ask. Individuals demand less proof for what they are inclined to believe is true than for what they doubt. When knowledge is applied in a political context the ideological component becomes more influential. Yet factual evidence which does not correspond with what is expected can call even deeply held beliefs into question. Perhaps the most exciting experience that a scholar can have is to encounter evidence that runs contrary to what he or she expected. While dogmatic or ideologically driven individuals may ignore such evidence and even suppress or distort it in order to uphold cherished beliefs, deliberate behaviour of this sort is almost always condemned. Most scientists encountering strongly contradictory evidence are likely to begin rethinking key issues. Contrary evidence may lead scientists and those who are aware of scientific findings to alter even deeply held political beliefs and commitments. In McNeill's terms, this involves forging a new and more appropriate mythistory.

The Ecology Movement

An example of the contemporary operation of mythistories is provided by the ecology movement. As we have already noted, concern with environmental degradation, the depletion of non-renewable resources, and the destructive effects of uncontrolled population increase were all introduced into sociocultural evolutionary studies in the context of the dystopian evolutionism of the 1960s and 1970s. The initial effect was to challenge the concept of change as an absolute good and to legitimize the *status quo* as a less dangerous alternative. Today, as major elements of the ecology movement, these same concepts have become part of a radical and far-reaching critique of the values and priorities of modern industrial societies. With the collapse of Marxism as a political and moral force, the ecology movement has emerged as the most powerful opposition to the dominant neo-conservative philosophy and its efforts to legitimate unrestricted economic growth and minimally regulated market economies. Because they examine long-term relations between human beings and the environment, evolutionary and archaeological studies have an important role to play in current political debates and the construction of present and future mythistories.

The ecology movement began in North America in the nineteenth century with lobbying to create national parks and protect endangered wild life. These activities were deeply coloured by a romantic view of nature. While the original ecology movement did not challenge the inevitability or desirability of sociocultural progress, it did deplore some of its consequences and tried to mitigate them. Today the ecology movement has become far more diversified and has split into a number of rival camps. These fractions share a concern with the impact that human exploitation has on the environment: in particular with the effects of an increasingly complex technology, a rapidly growing global population, and the impacts of transnational corporations on the ecosystem. Self-styled ecological realists seek to pinpoint specific problems and resolve them. They believe that the mobilization of public opinion is reversing some of the most dangerous trends and that through public education it may be possible to stave off ecological disasters such as global food shortages and the collapse of the biosphere (Easterbrook, 1995).

Deep ecologists deplore the utilitarian ethics that they see underlying the environmental management for human purposes being advocated by ecological realists. They trace the present ecological crisis back to an alleged Judaeo-Christian lack of respect for nature, as embodied in the Old Testament claim that God gave human beings power to use other living things for their own purposes (White, 1967; cf. Hughes, 1975: 141–6). They interpret this as an incitement to the ecological mismanagement that has characterized the societies of the Middle East and Europe from the beginnings of food production to the present. In its place deep ecologists expound a spirituality that embraces a pantheistic view of nature. They deny a bifurcation between the human and non-human realms and seek to revive a view that treats all facets of nature as elements of a great circle of life that is devoid of hierarchies or polarities. Eschewing a human-centred view of nature and a mechanistic world view, deep ecology traces its roots to nature worship, animism, pantheism, wicca, Native American religious beliefs, Taoism and other eastern philosophies, and the Gaia hypothesis. It seeks to engender a reverence for nature and a sense of oneness with nature as a catalyst for social action (Devall and Sessions, 1985; Drengson and Inoue, 1995; Pepper, 1993; Taylor, 1995; Zimmerman, 1994).

To counter the threat that the ecology movement poses to

unregulated economic growth, neo-conservatives have launched an anti-ecological crusade (Rowell, 1996). Its proponents argue that threats to the ecosystem are being grossly exaggerated in order to frighten people and that there is little reason to credit claims about global warming and other supposedly impending large-scale ecological disasters. They maintain that humans have had negative impacts on the environment from early times; citing evidence that Palaeolithic hunter–gatherers appear to have played a role in the extinction of numerous large animal species (Ehrlich and Erhlich, 1981; Martin and Klein, 1982; Nitecki, 1984). They also claim that advocates of ecological disaster underestimate the long-term resilience of the natural environment and exaggerate the impact of technology on that environment in their efforts to fetter economic growth and subject human lives to unnecessary and debilitating planning and control by left-wing ideologues (North, 1995; Ridley, 1995; cf. Bloom, 1995).

On a more philosophical level, the ecology movement is being attacked on the ground that those now living cannot be asked to sacrifice their personal interests for the hypothetical benefit of future generations (Beckerman, 1995). This is an extreme expression of the *laissez-faire* view that society works best if each individual attends to his or her own business and does not attempt to regulate the conduct of others. Such a view is the precise opposite of oft-repeated Native American claims that in their traditional societies no decision was ever implemented without first considering what its impact would be for at least seven generations. It is also an interesting counterpart to the eighteenth-century Enlightenment assertion that hereditary privileges, such as those derived from feudal land tenure, should no longer be allowed to fetter the economic activities of living individuals. This objection eventually led to the almost total absorption of land into an expanding market economy (Polanyi, 1944).

Deep ecology has been accused by its critics of being a refuge for nostalgic anti-progressivism as well as for leftist agitators. Its proposals to grant rights to animals, trees, and rocks have been characterized as being derived from a profoundly anti-humanistic doctrine that shares much in common with the totalitarian tendencies of German romanticism, which are alleged to have deeply influenced the Nazi movement (Branwell, 1989, 1994). It appears profoundly retrograde for some deep ecologists to be urging people to readopt an animist view of the universe, when a clear understanding of the differences between the human, the natural, and the physical realms is among the most far-reaching and empowering sets of insights that have been achieved in the course of human history. Nevertheless, the neo-conservative critique raises interesting questions about its own internal consistency when its advocates argue simultaneously that a human-centred view of the universe is essential to preserve democracy and that technological progress and monetary profit should be put ahead of individual well-being. Those who advocate personal happiness as the supreme good can hardly use the argument that present pain leads to general long-term gain as a basis for demanding sacrifices that will benefit only future generations.

By becoming involved in political action, ecology is being drawn into debates that penetrate to the core of Western philosophy and values. These debates are essentially irresolvable, since the key concepts are polyvalent and can be used to support many different and even diametrically opposed viewpoints. Philosophical positions seem in many respects to resemble natural languages. Anything can be said in any language even if the difficulties involved in expressing particular ideas differ from one language to another. Likewise, with enough ingenuity individual philosophies can be construed to support almost any political agenda.

The ideological implications of ecological research have also given rise to factual debates that have major political implications. In recent years it has been argued that Native American forms of agriculture, that are believed to have been inspired by a respect for nature lacking in the Judaeo-Christian tradition, were environmentally far less destructive than either earlier European or modern Western forms. A key example is the Valley of Mexico, which sustained a very high population density without major ecological degradation prior to the arrival of the Spanish conquerors in 1519 (Sanders et al., 1979). Despite a massive reduction of population as a result of outbreaks of hitherto unknown European diseases, the introduction by the Spanish of goats and European seed crops that were grown on the hitherto uncultivated higher flanks of the valley led to deforestation and massive soil erosion that, within twenty years, had silted up the highly productive lake system in the centre of the valley (Hassig, 1985: 207–8). It is argued that the ancient Mexican system of *chinampa* (raised-field) agriculture was both highly productive and ecologically friendly. This form of agriculture involved creating long, narrow fields in shallow, swampy areas by piling up soil and using organic fertilizer to maintain a high level of fertility. If well managed, such plots could produce multiple crops each year over a very long period. Today efforts are being made to revive this form of agriculture, both to understand better how it worked and as an alternative to conventional Western-style farming. The latter involves clearing large areas of rainforest to produce beef, as well as the privatization of land for large-scale commercial agriculture. Chinampa agriculture is highly labour-intensive and best managed by individual farmers or farm families.

This attempt to revive pre-Hispanic forms of agriculture is interpreted by many as a criticism of current economic trends in Mexico and therefore as being part of the political resistance to what currently passes for modernization in that country. Not surprisingly, considerable efforts are being made to discredit the idea that aboriginal North Americans were ever natural ecologists. Wide coverage has been given in the press and on television to archaeological findings at Copan and other lowland Maya sites which indicate that the Late Classic period (AD 600–900) was characterized by rapidly growing population densities and by soil erosion which might have led to the collapse of civilization in that area (Fash, 1991). Similar attention has been paid to evidence that soil erosion around Lake Patzcuaro in west-central Mexico might have been almost as severe prior to the arrival of the Spanish as it was after (O'Hara et al., 1993). The reports of this finding failed to note that, as a result of European diseases, the population of that area was much lower after 1519 than it had been previously. In both cases discoveries that appear to undermine the idea that native people were natural ecologists have been widely disseminated (Deneven, 1992). Whether the original researchers were merely anxious to emphasize evidence that appeared to refute a widely held view or were ideologically committed to the New Right is irrelevant to the attention that has been accorded to these findings in the popular press (Stevens, 1993). They are clearly being used by the media to refute suggestions that there are agricultural practices that are ecologically more friendly than those currently being pursued.

On the other hand, in addition to the question of whether traditional forms of agriculture could sustain a high level of productivity without inflicting as much ecological damage as modern forms of agriculture have done, it must be enquired whether these ancient methods of agriculture could sustain Mexico's present overall level of population or the lifestyle to which modern Mexicans aspire. It must also be asked whether, if the aboriginal Mexicans made less of a negative impact on their environment (at least in some areas), this was because their belief system inculcated a respect for nature or because they lacked sheep and goats that destroyed natural vegetation and draft animals whose ploughing exposed large areas to soil erosion. The absence of these domesticated animals clearly encouraged less destructive, but more labour-intensive, forms of cultivation. Having no direct way to get inside the minds of ancient Mexican farmers, this is a hard question to answer.

The Limits of Objectivity

In chapter 10, we argued that studies of past and present changes can be relevant for understanding future trends. Such findings in turn are relevant for constructing and deconstructing the myth-istories that guide modern political life. For research to be of maximum utility, however, it is necessary that it should be inspired by as many different viewpoints and commitments as possible, including ones that are controversial and unpopular. Studies relating to the biological basis of gender may be anathema to those who do not believe that such differences exist and who, not without

234 EVOLUTION AND THE FUTURE

reason, fear that fraudulent or biased findings may be used to disadvantage women economically and politically. Those who deny false memory syndrome often claim that any efforts to prove that it exists slander victims of childhood sexual abuse. Those who hold the opposite view maintain that the inability of therapists to distinguish real memories from false ones leads to the wrongful victimization of individuals and families in many cases where, without additional substantiating evidence, childhood sexual abuse is alleged to have occurred (Bunge, 1996: 206). If scientific investigations are to help resolve social issues, and not simply generate propaganda to promote causes, the public must be prepared to support research inspired by opposing views of the same issues, while ensuring that all findings are subjected to rigorous review and evaluation. The greatest danger is the domination of research by preconceived notions of political correctness or by self-serving agendas established by business people, civil servants, governments, and other sectoral interests.

The problems posed by single interests can be controlled to a considerable degree by public funding of research and peer evaluation of proposals. Political correctness requires more vigilance, because much of it is related to academic politics. Researchers must cultivate among themselves an enduring commitment to open research, broadly based peer reviews of findings, and rigorous counteracting of the dissemination of erroneous or inadequately substantiated findings for political purposes. Ecological issues cannot be divorced from political programmes. Yet, as ecological processes are better understood, that understanding provides a firmer basis on which the effectiveness of specific conservation programmes can be evaluated. While this does not provide a substitute for political action, or guarantee that scientific information will not be misused, it does offer a more informed basis for political activity.

The direct influence of scientific findings on political actions must not, however, be overstated. Sometimes the same understanding of reality can be interpreted in diametrically opposed ways. Darwin's notion of group selection was construed by anarchists, such as Kropotkin, as evidence of the general benefits of co-operation, but by racists and nationalists as an argument endorsing ethnic conflict. Likewise, both Karl Marx and the Austrian economist Friedrich von Hayek, on the basis of the same ethnographic evidence, agreed that early human societies shared a redistributive ethic. This meant that, however much each individual produced, everyone received an equitable share of the necessities of life. Both economists also agreed that this aspect of early human behaviour was relevant for understanding modern politics. Marx, however, saw it as proof that all human beings could become sharing and concerned for others if they were freed from the competitive ethos of capitalist societies. Hayek believed that early man was over-socialized and hence, as a result of cultural conditioning, was opposed to all that was innovative, creative, and progressive. He argued that respect for abstract rules had to replace both a love of fellow humans and a sense of shared purpose, if mutually beneficial progress were to result from the ruthless pursuit of individual self-interest. While Marx sought the return, at a much higher technological level, of societies governed by humanity's true nature, Hayek believed that primordial nature had to be transcended for life to be worth living (Gellner, 1988: 27–9). Ironically, it now appears that Marx's and Hayek's shared understanding of early human behaviour was factually incorrect, since in small-scale societies both economic and social equality exist only because they are supported by highly effective systems of positive and negative rewards enforced by public opinion. This suggests that they are not simply reflections of human nature. Nevertheless, because of their very different values and aspirations, Marx and Hayek interpreted what they both agreed was true very differently.

Finally, where significant material interests are involved, political agendas tend to be pursued regardless of the relevant evidence or how it is understood. In the 1820s there was a major debate among Euro-Americans about whether or not North American Indians were capable of adopting a Euro-American style of life. East coast intellectuals, who continued to be influenced by the Enlightenment ideals that had played a major role in the American Revolution, argued that they could and that it would be best for them to be absorbed as quickly as possible into the mainstream of

Euro-American society. Frontier ideologues maintained that Indians could not be civilized and therefore should be forced to shift westward to make way for Euro-American settlers. Both agreed, however, that the Indians should have their land taken from them. Those who advocated civilizing them believed that restricting their land base would force aboriginal people to give up hunting and become Euro-American-style farmers; those who maintained they could not be civilized argued that, if Euro-American settlement were to proceed, it was best for Indians to retreat farther west, where at least for the near future they might continue to maintain their primitive style of life. Otherwise, it was inevitable that they would die out as soon as Euro-American settlement deprived them of their hunting territories. The progress that many native groups living in the eastern United States had made in adjusting to the lifestyle of their Euro-American neighbours and in adopting many elements of that lifestyle did not prevent their deportation to Oklahoma as a growing number of Euro-Americans coveted their national territories east of the Mississippi River (Green, 1996: 510–33).

History suggests that, although ideas and values more often serve to condone than to modify human conduct, they can at critical points in human history play a significant role in determining what happens. Neo-conservative philosophy, combined with major technological changes, provides a spectacular example of this. It would appear that for the most part new (or what appear to be new) ideas are most effective when they are perceived as promoting material and political self-interest. Fortunately, self-interest, even when narrowly construed in terms of material gain, is not a given. It can be defined in relation to broader or narrower frameworks, and to shorter or longer time perspectives (Hardin, 1968). Which of these definitions is accepted may not be determined entirely by cultural presuppositions but be susceptible to reasoned discussion.

One, perhaps fragile, hope of bringing about progressive change is to attempt to redefine the perceived self-interest of dominant groups so that it becomes more congruent with the general well-being of society. In a world in which any chance of capitalism collapsing in the near future, except perhaps as a result of its own excesses, seems exceedingly remote, such a possibility of reform deserves careful consideration. This sort of change involves the propagation of new mythistories. While the prevailing neo-conservative ideology was developed alongside the growth of new communications technologies that have been used very success-fully to weaken the regulatory powers of nation states over their economies, there is no evidence that a neo-conservative society is necessarily inherent in, or permanently associated with, that par-ticular technology. Moreover, even if constant technological change is not characteristic of all, or even most, societies that have ever existed, it has emerged as a dominant feature of the most complex societies of recent millennia and especially of the last 500 years. That development is not a figment of evolutionists' imagina-tions but something that has come to affect the daily life of everyone living on this planet. To deny this development is to ignore some of the key material and ethical questions of our era, which have to do with changes within societies and in relations of exploitation, emulation, and colonization among societies. Under these circumstances no form of ideology, including a neo-conservative one, is likely to persist unchallenged for very long.

Ernest Gellner (1988: 140–2) saw a key feature in the creation of the modern world being a move away from universally valid Platonic archetypes as prescriptions for human conduct toward a view of social reality and morality that is constantly changing and renewing itself. This defines the essence of McNeill's mythistory. There are, of course, behavioural constants that relate to human beings as biological organisms which appear to change very slowly, if at all. Yet, when it comes to modern, technologically evolved societies, social and ideological change is the norm. Seemingly fixed systems of morality and ethics are only possible in societies where change takes place so slowly that it cannot be perceived or, under pathological conditions, among groups that feel so threat-ened by change that they seek to deny its reality. Otherwise, it is no longer possible except in the broadest terms to view issues of public order, personal freedom, and cultural diversity in absolute terms. It is necessary to determine in what varied ways practices can change as material conditions alter. To envisage a more satisfactory future we need new mythistories.

The general outline of what has been happening is clear. Even early band and tribal societies found themselves parts of larger networks of social and economic relations that were vital for mutual survival. Since then, as Alexander Lesser observed long ago, 'the interdependence and interaction of human groups . . . has been developing in extent and in intensity throughout human history' (Mintz, 1985: 141). Only recently, however, through a process of internationalization have human beings begun to perceive themselves as inhabiting a small, interactive, and ecologically vulnerable planet. Yet effective interdependence is extremely difficult to realize because of deeply entrenched economic, political, and cultural interests that vary significantly from one place and one time to another. An overall trend is accompanied and obscured in significant ways by vast local and regional variation.

The Challenge of Rapid Change

Today technological change is occurring faster than ever before and its impacts on society and the environment are now worldwide. The growing interdependence of different regions allows specific groups ever less freedom to follow their own course of development. At the same time, unprecedented intercontinental migration resulting from political oppression, individuals' searching for employment, and a desire for greater personal fulfilment is transforming major cities and even smaller centres in the developed world into microcosms of the ethnic and cultural mosaic that formerly was found dispersed around the planet. Many migrants are clinging to their ethnic identities in an effort to preserve meaning in their lives in a rapidly changing world. The resulting concentration of cultural diversity is a source of economic as well as cultural enrichment for Western societies.

In order to cope with these social changes, new values are emerging. The melting pot, which once sought to assimilate immigrants, has in an age of transnational economies given way to a multiculturalism that stresses the social and cultural benefits to be derived from cultural pluralism and diversity. Yet multiculturalism poses its own problems. Its limits are tested when, in the name of cultural relativism, religious tolerance, and even gender equity, societies are called upon to accept female circumcision as a religious rite, just as they already do infant male circumcision, or when it is argued that some teenage boys should be allowed to carry knives in schools for religious reasons or because doing so is a sign of manhood in their parents' culture. How does a modern, multicultural society respond to a family that threatens a teenage daughter with physical violence if she insists on dating a young man from a different ethnic or religious group?

There are few obvious and simple solutions to most of these problems. At this point the individual rights championed by the Enlightenment clash with the collective rights promoted by romanticism. Yet the notion of universal rights requires that, in the interest of sustaining subordinate collectivities, no individual should ever be obliged against their will to surrender any of the general freedoms to choose that are rightfully theirs as human beings. Recently, apologists for despotic regimes in East and South-East Asia have sought to circumvent this argument by claiming that individual liberty is a culturally specific belief of Western civilization that has no counterpart in traditional Asian cultures, which are constructed around notions of collective identity. The tenacious struggle for individual rights and freedoms being waged within these same countries effectively gives the lie to such self-serving propaganda. It also reveals how Western concepts of relativism are being manipulated in the interests of tyranny and draws attention to a dark side of romanticism.

At the same time, in order for societies to function, the members of every society must agree about some basic rules. A modern industrial society cannot perform well if it becomes rigidly fractionated into ethnic or religious enclaves that put their own interests ahead of those of the society as a whole. The existence of such divisions inhibits the social and economic flexibility that is necessary for such societies to respond optimally to technological and economic opportunities. There is also no greater danger to the survival of a democratic society than the attenuation of the ideal that every citizen shares the same rights and responsibilities.

In a Durkheimian fashion, increasing economic interdependence and a growing division of labour make it possible for a well-integrated society to cope with increasing cultural diversity and a larger number of idiosyncratic lifestyles, which should mean that individuals have greater personal choice about how to live. Modern multicultural societies are slowly and painfully groping towards a better understanding of how to protect individual freedom of choice while respecting cultural diversity. This appears to require cultural diversity both within and between societies to evolve in the direction of freely chosen and often multiple cultural identities rather than remaining associated with regionalized or ghettoized minorities. Nevertheless, economic and social insecurities and disparities breed atavistic animosities that make this more personalized approach to cultural diversity very difficult. If the general process of creating multicultural societies is easy to comprehend, the problems involved in actually producing such societies are very great.

Because of the increasing tempo of cultural change and the greater degree of interconnection linking all human societies together, there is also ever-diminishing scope for trial and error. During the Palaeolithic period, if one band of hominids made an adaptively fatal decision or fell victim to some irresistible natural catastrophe, similar hominid societies would survive elsewhere. If one early civilization collapsed, other civilizations would survive or could be expected to arise in other regions. Today, however, as human beings become enmeshed in a single interactive economic and political system and technology affects the entire biosphere, often in little-understood ways, the future of all humanity increasingly depends on the decisions that each and every human society makes (Dunn, 1993: 133–4).

Planning: Its Perils and Possibilities

Under these circumstances, it becomes ridiculous to argue that planning is not needed. The expansion of the brain's powers of foresight and calculation is the major characteristic of hominid biological evolution. It is the capacity to envisage alternative courses of action and to anticipate their consequences that has resulted in human beings becoming in Darwinian terms the most successful animal on earth. Planning is also what neo-conservatives argue individuals must do if they are to be successful. Indeed, as capitalist society becomes more complex, successful investment, like life in general, depends increasingly on planning, co-operation, trust, and teamwork rather than on competition. To claim that each person pursuing his or her own goals will automatically produce the greatest good for the greatest number is as mystical as the vulgar Marxist certainty that humanity is inevitably evolving towards a classless paradise. It is also a far more dangerous concept. To ensure a better future, it is necessary to forecast the results of collective as well of individual behaviour. To manage an economy we need not only microeconomics, as some radical neo-conservatives maintain (Brenner, 1994), but also macroeconomics.

The development of a global economy and of global ecological problems indicates the urgent need for authority at that level to monitor change, estimate impacts, and regulate technological and economic development in the common interest of all human beings. I am not proposing a planned world economy, since individual initiative and choice are effective devices for regulating many aspects of production and distribution. But it is necessary to subject worldwide economic activity to political control at least to the same extent as it was regulated within nation states prior to transnational companies breaking free of such restrictions. Such control included curbing monopolies, monitoring environmental impacts, limiting working hours, regulating health and safety in the workplace, establishing environmental standards, and ensuring that businesses contributed their fair share of the cost of maintaining essential social institutions, infrastructure, and amenities. Many of these obligations may require businesses and individuals to pay higher rates of taxes in times of prosperity to buffer periods of lower economic activity. An international agency could also establish trade and pricing policies that would help to eliminate the worst economic disparities between richer and poorer countries.

Planning at any level is notoriously difficult. As societies grow increasingly complex they also grow increasingly opaque, in the sense that it becomes harder for well-intentioned people to estimate what the consequences of their actions will be and correspondingly easier for greedy and exploitative people to disguise what they are doing (Baum, 1996: 26). Planning is therefore subject to gross distortions of perspective, corruption, and the favouring of the interests of the powerful. This is especially so if it is carried out mainly at higher levels and by self-styled experts. The development literature is filled with horror stories about the disastrous consequences of schemes that have been imposed on indigenous peoples by poorly informed foreign development agencies. Ecological and economic planning needs to have significant inputs from the local level, and its effects must be monitored and significantly controlled at that level if it is to be effective (Brohman, 1996; Chambers, 1983; Salisbury, 1986). In order to balance local and global concerns, planning has to be carried out in an interdigitated fashion at the local, national, regional, and world levels. In most parts of the world, this would require a reallocation of current powers downward to the regional and local levels and upward to the global one.

The development of transnational economies and communication systems has encouraged the devolution of decision-making from the national to the regional level of government. Yet this has been accompanied by the making of all political decisions in a context so dominated by neo-conservative economic policies, and by transnational financial institutions, such as the International Monetary Fund, exerting controls over governments, that overall political activity tends to be curtailed rather than enhanced as a result of decentralization (Luttwak, 1996). The slashing of national welfare programmes has encouraged some people with low incomes, often in co-operation with community organizations, to work together to create micro-systems that cater for local needs, such as common kitchens, jointly run stores, ventures to repair homes, efforts to grow food in backyards or on allotments, loan associations, and worker-owned small businesses (Baum, 1996: 57–8). These associations may play an important role in helping to create new forms of political power that will grow from the bottom up and eventually may promote more effective political action at the local level. On the global level, however, they are likely to provide no more than a first-aid solution for massive haemorrhaging (Welch, 1996a).

The creation of a global society capable of subjecting transnational economic activity to a greater degree of political control than at present would bring an end to the unforeseen impacts on sociocultural evolution that result from intersocietal competition. That in turn would limit to a considerable degree the cultural equivalent of natural selection, which until now has played such an important role in determining the overall direction of social and cultural change. It would, however, as the physiocrats realized long ago, not entirely eliminate it. While humans, individually and collectively, have the power to choose, some choices produce better results than others, and some result in disaster. Even in the absence of uncontrolled intersocietal competition, such selection would continue to pass its impersonal judgement on policy decisions. Nevertheless, overall sociocultural evolution would be directed to a greater degree by informed decisions made by human beings acting in a public capacity at every level from the local to the global. Thus it would be governed to a greater extent than in the past by moral forces and to a lesser degree by forces that lie beyond human understanding and control. The possibility of building societies that are just and humane would be enhanced and human beings would be endowed with unprecedented collective as well as individual freedom (Baum, 1996: 28–9).

Since 1945, there have been a number of successful efforts to build larger economic and political blocs. Most such efforts, however, have been hindered or subverted by the resistance of existing national governments to surrendering their powers. During this same period, nationalism has torn apart a number of multiethnic states, such as Pakistan, the Soviet Union, Czechoslovakia, and Yugoslavia, while regional and elite interests have defeated movements to unite states that share the same ethnic identity (the Arab Nation). It is evident that political unity has been created at ever higher levels throughout history, but there are few examples of the leaders of existing states surrendering significant powers unless they have been forced to do so (Carneiro, 1970b). Despite passionate professions of enthusiasm for German and Italian unity in the nineteenth century, it is unlikely that either of these states would have been unified had it not been for the Prussian and the Piedmontese armies.

At present the institutions that have the most to lose from the development of higher levels of political control are the transnational corporations. These corporations can exploit nationalism to try to prevent or delay the emergence of a political force capable of dealing with them on equal terms. The only likelihood that their attitude would change would be if nationalist rivalries were to undermine order within and between enough states that the resulting chaos threatened the profitability of a transnational economy. While national governments and economic associations such as the European Union seek to preserve their powers of self-government, these powers have become increasingly limited by comparison with those of transnational companies. This has compelled most national governments to reduce corporation taxes and to seek to control the national deficits that result from doing this by cutting back on social services.

The emergence of higher-level political controls is also opposed by the neo-conservative ideology which provides the current justification for *laissez-faire* economic policy. Patricia Marchak (1991) has documented how in the 1950s, neo-conservatives, supported by lavish, tax-free funding from private foundations, began to seize the initiative in ensuring that important technological innovations, especially in the fields of data processing and communications, would be used to reshape society in accordance with their *laissez-faire* ideology. They argued that, because human societies were complex and individual tastes varied, it was impossible to predict individual requirements; hence these were best supplied by an unregulated market economy and individual decision-making. A well-orchestrated and sustained assault on rival positions soon compelled complacent liberals and social democrats to debate economic issues on terms set by the neo-conservatives. This framework became so powerful and all-encompassing that until recently it seemed impossible to discuss social issues on any other grounds. The neo-conservative hegemony has materially assisted transnational companies to undermine the regulatory powers of even large nation states, while impeding the development of effective international agencies that might monitor and control their activities in the public interest. The consequence has been a flow of power away from elected governments into the hands of corporations that are answerable only to their shareholders (Sassen, 1996).

Neo-conservatism is promoted as an academic discourse not only by the accolades that it receives from politicians and business people. It is also supported by lavish funding for research from rich individuals, foundations, and companies, privately endowed teaching positions in universities, and the numerous awards and prizes that its promoters have established to honour and advantage young researchers who promote this point of view (Lemann, 1997). These tangible rewards have been of no small importance in helping neo-conservative ideas to achieve and retain their hegemonic status. Unhappily, the success of neo-conservatism has driven many of its opponents to adopt a postmodern viewpoint as a form of self-defence. Yet, by denying the possibility of creating objective knowledge about human behaviour that could directly challenge neo-conservative claims, these critics have proclaimed their inability to create a credible alternative to neo-conservatism. In so far as it embraces extreme relativism, postmodernism or any other philosophy admits its inability to influence the course of history. In this fashion, even postmodernists who disapprove of neo-conservatism become unwittingly complicitous in promoting it, and postmodernism, willingly or unwillingly, becomes the culture of transnational capitalism. Neo-conservatism must be met head on and its intellectual flaws revealed in an empirical and convincing fashion.

If planning to avoid ecological problems is to be effective, it requires democratic governments and an educated and responsible citizenry everywhere. The ecological disasters that have permanently damaged huge areas of the former Soviet Union and Eastern Europe show what can happen when citizens are unable to protest the abuse of their local environment. The development of a global level of government without local surveillance and decision-making would amplify rather than resolve current problems, since it would merely complement economic domination by transnational corporations with more political decision-making from the top. Nothing would be done to ensure that local knowledge and concerns were duly considered as part of the overall regulation of economic activities. Likewise, even if decisions about development are made democratically by the entire populations of hard-pressed areas, optimizing short-term economic gains may still count for more than resolving ecological problems. Poor regions, especially ones with less-educated populations and less access to international media, would remain at greater risk of falling victim to ecologically damaging forms of economic exploitation. The protection of the global ecosystem requires the achieving of enough regional economic equality to encourage local willingness to deal with ecological problems. This alone will produce sufficient co-operation and goodwill among people living in different regions to moderate the destructive competition that results from the desperation caused by poverty and the efforts of local as well as foreign elites to profit from the exploitation of backward regions. Such equality would also reduce growing north–south antagonism and help to curb the rising opposition between those who are profiting from rapid economic change and those who are being harmed by it. It is unreasonable to expect third-world countries to limit their economic development or increase its cost to themselves in order to help solve global ecological problems without proportional sacrifices being made by industrialized nations.

Creating New Identities

It is also important to promote a sense of identity transcending that of nationality, language, race, or religion so that it becomes possible for individuals to engage in a positive fashion with the whole of humanity. Over the course of human history, there has been a broadening sense of identity and mutual responsibility from bands and tribes to nations and even larger religious and ethnic groupings. This suggests that a further expansion of identity and loyalty is not impossible. For a very long period human brotherhood has been stressed as a desirable goal by various international religions, but always on the precondition that individuals subscribe to a specific code of beliefs and practices. As yet, however, a sense of pan-human identity has remained more a philosophical concept than a political one.

The expansion of loyalties to embrace humanity as a whole would not extinguish narrower loyalties. Indeed, by helping national governments to curb the anti-social activities of transnational companies, expanded loyalties might increase, rather than diminish, the respect that citizens feel for national governments. However, a pan-human loyalty would be unique in one respect: it would not leave an outgroup. Hence the unity of this group could not be reinforced by pitting its members against one or more other groups. This is a type of identity-building that has only been experienced in past times by relatively small human communities living on remote islands or in the high Arctic; places so remote geographically that contact with other human groups was lost for long periods.

Bringing about a change of this sort would require an immense intellectual transformation. In recent decades the Western mass media have successfully promoted neo-conservative ideas in every part of the world. They also played a major role in undermining the Soviet Union and bringing an end to its attempts to build an alternative type of industrial society. One can scarcely expect the information media, which are owned and controlled to an ever-increasing extent by transnational companies, to promote ideas that run contrary to the short-term interests of their owners. What is being promoted today as a definition of what is pan-human is dictated by the United States' entertainment industry and its largely neo-conservative ideology. To oppose the neo-conservative agenda, it is necessary to wrest control of the mass media from the neo-conservatives and ensure that there is a physical basis for articulating the discussion of alternative ideas on a worldwide scale. Doing this will require at least as much imagination and resourcefulness in utilizing changing technologies as neo-conservatives have displayed since the 1950s. I do not know how this may be accomplished or of anyone who does. It may involve the building of alliances among groups that are dissatisfied with various aspects of the neo-conservative agenda but do not yet appear to be likely allies.

Controlling Development

Those who regard rapid and uncontrolled technological change as being both ecologically and socially dangerous must consider the alternatives. It is suggested that halting further technological development might avert future ecological and social dangers, either by preserving the existing political order or by permitting the creation of a new, utopian one. Neither of these anticipations appears justified. Even if technological change were stopped, alterations in population density or in the natural environment would continue to bring about social change. In small-scale societies, such changes tend to be limited under normal circumstances, but in complex societies the impact of demographic or ecological changes tends to be far-reaching. No matter how much people in such societies may seek to avoid change, it appears to be inevitable. The only questions are whether changes will lead to the expansion or the contraction and possible collapse of societies and whether, if there is expansion, it will benefit the few or the many.

Anthropological studies indicate that some degree of technological change and economic expansion may be required to maintain an open society (Goodenough, 1972). It appears that an economically stable society, even a universal one that faced no external competition, would tend to become more hierarchical and repressive. Those in privileged positions would seek to pass their privileges on to their children and resort to increasing economic and political manipulation to assure this happened. This sort of behaviour has already become evident in Western societies as their rate of economic growth has slowed from its post-Second World War high point. Neo-conservatism is in part an ideology that seeks to free the wealthy and powerful from having to support a public welfare system so that they can devote their efforts to protecting their own economic interests and those of their families in a period of overall economic stagnation (Krieger, 1986). It has also been demonstrated that maintaining a society in a steady state or coping with economic deterioration and growing economic inequality requires a higher level of repressive control and regimentation (Smith, 1986).

A more desirable option appears to be sustained, monitored, and regulated development. Monitoring and controlling development would enable the rate of change to be slowed so that the long-term ecological, medical, and social impacts of technological innovation could be measured (Tenner, 1996). It would also place major restrictions on free enterprise by re-embedding economic activities into a broader social, political, and cultural context. Were sustained development to become impossible for ecological or other reasons, such as the failure to develop new, clean, and inexpensive sources of energy, more rather than less social planning would be required to try to counteract the worst social consequences of a static or contracting economy. Population control seems to be one of the most promising means for regulating relations between societies and their natural environments. By lowering population levels, a point could be reached at which sustained development might be resumed. Population control sometimes involves considerable intervention by governments or societies in the most intimate human relations, as is currently happening in China. Yet experience in more developed countries suggests that in these societies individual assessments of the high economic costs and low economic benefits of having children may be more than sufficient to reduce the birthrate to manageable, or even unduly low, levels.

Today human beings face a double dilemma. History suggests that technological change may be essential for individual and collective self-fulfilment of the kind we have come to expect, even if it does not have to occur at the hypertrophied rates currently being experienced in the most buoyant industrial economies. Yet there is also a growing need for far-reaching planning and international supervision and control, if the destructive ecological and social

effects of technology are to be kept within limits that are acceptable
to a majority of human beings. Bureaucracies, as we have already
noted, often tend to become self-serving and exploitative, and this
exacerbates problems of managing change. *Laissez-faire* means that
most people do not know who is manipulating the economy, and as
a result individual, short-term interests determine most outcomes.
Neither option protects the environment or ensures a just, and
ultimately sustainable, society.

What is needed is planning that takes place in an open, trans-
parent, and democratic fashion at every level from the smallest unit
of settlement, a house or farm, to the world system. Even if it is
objected that such planning is cumbersome and inefficient in the
short term, it is surely preferable to irreversible ecological disasters
and the creation of societies in which the public sector is starved of
resources to such an extent that daily life becomes intolerable even
to those who are wealthy and powerful enough to isolate themselves from the social chaos which their own selfishness and social
irresponsibility have led them to create. As technologies and societies grow more complex and change occurs more quickly, the
dangers of social pathologies leading to systemic breakdown
increase. Furthermore, while a diminution in types of societies
reduces the pressure of destructive intersocietal competition, it
increases the dangers of social pathologies spreading unchecked
through all societies. While planning may, as many experts fear,
prove inadequate to counteract these pathologies, it is clearly
better than doing nothing.

An Alternative to Relativism?

Sociocultural evolutionists stand accused of denigrating non-
Western cultures by characterizing them not merely as different
from their own but as primitive types that have been rendered
obsolete by Western societies. Yet, ever since the eighteenth century, some evolutionists, as well as anti-evolutionists, have looked
to small-scale societies to provide models of alternative and better
lifestyles than are found in hierarchical ones. While Marx and

Engels saw what they interpreted as the freeing of human beings from direct control by natural forces as being highly desirable, in their scheme of sociocultural evolution they viewed hunter–gatherer societies as proof that exploitation was not inherent in human societies. Today anthropologists have been humbled by their growing awareness of the complex understandings that aboriginal peoples have of their environments, the skill with which – despite limited material resources – they have adapted to changing natural conditions, and their ability to exploit ecosystems in ways that tend to preserve rather than destroy these systems. The respect that many indigenous peoples have for nature, once dismissed as primitive animism, is now seen, despite its religious mode of encoding knowledge, to have significant features in common with a modern ecosystemic view of humanity's relations to the natural world.

It is also recognized that the experience of every human group is precious because it represents a facet of the totality of human experience. The knowledge possessed by hunter–gatherers is relevant to the needs of all other societies, however technologically advanced they may be. The exploitation of native peoples continues, especially in terms of resource deprivation, but there is increasing public recognition that this is unjust. It is also increasingly acknowledged, though more often in principle than in practice, that indigenous societies have rights to their land and to refuse its use for purposes of which they do not approve.

In recent years many social anthropologists and archaeologists have become involved in helping indigenous peoples to strengthen or reinvent their traditional cultures in the hope that this will assist them to resist colonial oppression and economic exploitation (Allen, 1992; Freidel et al., 1993). Yet, however much these well-wishers may seek to create conditions that would allow aboriginal peoples to cultivate their traditional ways of life and permit cultural variety to flourish on a global scale, no outsider has the right to consign other human beings to ethnological zoos. Native people as individuals must have the political and economic freedom to choose whatever lifestyle they find most fulfilling. They should be able to decide whether they wish to follow a traditional way of life, integrate into larger industrial societies, or pursue a middle course. For the decision to integrate to be morally valid, however, viable alternatives must be available and the choice must be free from coercion by a dominant society or transnational economic interests.

Another form of cultural diversity is no less seriously threatened, although it appears to have fewer defenders. Within the general context of democracy and individual rights, there is a need for variety in types of social, political, and economic organization in order to determine which work best in the long run and provide the greatest human satisfaction. Making criminal law a prerogative of state jurisdictions within the United States has been lauded on the grounds that it permits experimentation with different forms of dealing with offences. At the present time it would be useful to have more empirical knowledge about how economies based on varying types of state intervention might compare with those run on strongly *laissez-faire* principles. Yet this sort of diversity is opposed out of short-term self-interest by transnational companies and on ideological grounds by neo-conservatives, who seem intent on suppressing all ideas that are different from their own. Intense economic and ideological pressure is brought to bear on small countries with strong social democratic traditions to compel them to abandon such practices. This linearization of behaviour, which seeks to impose uniformity on all societies, surely poses a long-term danger, in so far as it forecloses on the possibility of experimenting with alternative ways of doing things. It would be highly counterproductive if, as the shortcomings of neo-conservatism become more evident, there were no clearly understood alternative forms of economic organization available to replace this intellectually bankrupt rehash of nineteenth-century *laissez-faire* liberalism.

Neo-conservatives tell us that we live in the best of times and should strive to be morally worthy of this privilege. If so, the best of times are haunted, even in the wealthiest countries, by widespread decline in personal purchasing power, economic dislocation, growing insecurity of employment, and increasing fears concerning impending social, political, and ecological disasters. Higher education and research are being starved of public funding in many countries, despite repeated assertions that the future of these countries depends on their creation of competitive information technologies. Government budget-cutting introduces increasing uncertainty into individual lives and makes it harder for governments to control economic fluctuations. Ironically, technological development, which only a few decades ago was regarded as full of promise for the future, now incites more fears than hopes. Planning is vilified, and the goals and desires of individuals are exempted from any counterbalancing responsibility towards the needs of society as a whole to an extent that was never imagined by the advocates of *laissez-faire* in previous centuries (Soros, 1997). Today it is widely maintained that a society cannot be regarded as democratic unless it embraces a free-enterprise economy. Yet the overthrow of democratically elected governments and their replacement by fascist juntas in countries such as Uruguay and Chile has been justified on the grounds that such action rescued these nations from the menace of socialism.

What is needed is a vigorous intellectual alternative to the equally problematical doctrines of neo-conservatism and centralized planning. This requires the rethinking of many of the basic concepts that have dominated Western civilization since at least the eighteenth century. In particular, it seems necessary to stop asking (and answering) questions in terms of counterproductive dichotomies, especially the weary dichotomies of rationalism and romanticism. Instead of opposing social planning and *laissez-faire*, it is more appropriate to enquire under what conditions each approach may have something to offer and how the two may be combined as part of a productive whole. Instead of opposing individual and collective rights, it should be asked how the interests of all human beings as individuals and as members of groups are best accommodated.

The Potential of an Evolutionary Perspective

In such rethinking, an evolutionary perspective has a vital role to play. Biological evolutionary theory suggests that the human capacity to conceptualize and reason symbolically, and to communicate ideas by means of language, expanded as a way for hominids to envisage and solve a growing range of problems. The resulting capacity to learn by precept, which allowed everyone to benefit from the experiences of individuals, proved to be a much more effective form of adaptation than did learning by means of personal trial and error. The ability to reason symbolically allows human beings to predict, at least to a limited degree, the consequences of alternative courses of action and to judge which of a number of different strategies might be either the most desirable, or the least undesirable, in terms of the goals they set for themselves and are capable of realizing. For most of human history, humanity's problem-solving capacities were used to deal with the ecological and social problems of small groups over short time frames. But, largely as a result of technological advances, in recent millennia they have been applied to coping with the more complex and longer-term problems of ever-larger societies. Larger and more complex societies require more co-operation, foresight, personal self-restraint, and an increasingly universal ethical awareness for their successful management. Where an ethical awareness of responsibilities to all human beings and to the planetary ecosystem is absent, the technological and organizational capacities of such societies can serve extremely destructive ends, including, in the twentieth century, the worst acts of mass murder carried out in human history.

Sociocultural change is not whatever people wish it to be. It is constrained by technology and by culturally transmitted knowledge which provides the guidance without which human beings would be unable to cope with their environment. Such knowledge also constitutes an inertia that opposes change, while at the same time it provides the concepts that are transformed as individuals and groups seek to cope with external challenges and to achieve new goals, often at the expense of other human beings. As a result of intersocietal competition, institutions and even whole societies that are less effectively organized than neighbouring ones can disappear.

Yet, if behavioural change is constrained by factors, both cultural and ecological, that are external to the individual, it is not wholly determined by these factors. Individuals and groups can adopt or reject new technologies, while the modification of social, political, and economic institutions involves complex negotiations among those involved. It is important to understand the role that constraints play when considering future options. But, even within the limits of these constraints, change can follow many different paths and produce many different results. Some of these outcomes are more viable than others, and of those that are viable, some arrangements serve the common good much more effectively than do others. No dogma is more pernicious in its effects, or more fallacious, than that of determinism. If human beings do not work together to shape their future, they may find they have no future, or at least not a future that is worthwhile.

Modern societies can choose to follow a policy of *laissez-faire*, which is rapidly increasing the gap between rich and poor, escalating conflict both within and among societies, and refusing to deal adequately with critical ecological challenges. Or they can try to produce more integrated societies that are based on a broad social consensus. These more integrated societies would, by redeploying the excessive profits of businesses that are currently being used to promote highly destructive forms of ecological exploitation, maintain an effective social safety net for all their citizens and ensure that economic change occurs within a framework that is both socially acceptable and ecologically responsible. The maintenance of such societies would require broadly based social planning in which the needs of individuals and of society as a whole both receive due recognition. Neo-conservatives argue that political intervention in the economy inhibits individual self-realization as well as overall economic growth. They also argue that inevitably the trickle-down effect of an expanding economy will benefit everybody. Their critics maintain that an uncontrolled economy will lead to increasingly dysfunctional gaps in wealth, education, and power among individuals and nations and that eventually unchecked competition and excessive ecological exploitation will lead capitalism to destroy itself, humanity, and the ecosystem. The most obvious shortcoming of neo-conservatism is that it is an unmodified replay of nineteenth-century liberalism, which in the late nineteenth and early twentieth centuries produced social strife, extreme nationalism, brutal totalitarian governments, racism, genocide, two world wars, and the deaths of many millions of human beings. Capitalism and democracy were rescued by social intervention in the form of Keynesian economics and the provision of social welfare, which was effective in maintaining social peace until a transnational economy outgrew the control of nation states.

Despite the seemingly inexorable expansion of technology from late Palaeolithic times to the present and of more complex societies at the expense of less developed ones, humanity does not have a guaranteed future. Instead, there is a range of possibilities extending from the total destruction of the ecosystem by thermonuclear warfare, or an insidious process such as the greenhouse effect, to a future filled with new, clean, and ecologically responsible technologies, which would make life truly better for an increasing number of human beings. Some neo-conservatives wish us to believe that history has already come to an end and that all that lies ahead is a future characterized by free enterprise. Given the evidence that new technologies encourage the development of new types of societies, this proposition is as unconvincing as the Marxist dream of a classless paradise. Social change will cease only when technology ceases to evolve and that would more likely result in social collapse or the evolution of highly repressive political regimes than in any kind of stability.

Despite the dreams of eighteenth- and nineteenth-century sociocultural evolutionists, there is no more scientific evidence of long-term purpose or transcendental inevitability in human history than in the natural world. Ian Simmons (1993: 186–7) has summarized biological evolution as: 'what happens next is simply a consequence of what happened immediately beforehand and what the local circumstances are'. Yet human beings have evolved the capacity to act purposefully, albeit on the basis of imperfect knowledge, both in their personal lives and together to achieve collective goals. That imposes on us a moral and intellectual responsibility for shaping the future not only of human societies but also of the natural world which produced and continues to sustain human beings. Thomas Huxley ([1893] 1947: 60–102) noted with Victorian assertiveness in his 1893 Romanes lecture on 'Evolution and Ethics' that 'the ethical progress of society depends not on imitating the cosmic process, still less in running away from it, but in combating it' ([1893] 1947: 82). Yet to what extent is it possible for human beings, as goal-driven organisms, to create a teleology for cultural evolution? Can we supplement our short-term personal goals with long-term social ones sufficiently to supply a conscious direction to the future course of our development? Can we hope to avoid economic and social collapse and to ensure the survival of the terrestrial ecosystem? And can we prevent industrial societies from mutating into some form of fascist tyranny as greedy and privileged individuals gradually surrender their own and everyone else's political freedom in order to protect their material possessions (Gellner, 1988: 231–2)?

It is evident that all too often humans are prepared uncritically to follow apparently successful role models and that as cultural change accelerates and communications systems increase in their rapidity and geographical range, the opportunity to test the efficacy of new forms of behaviour declines. The appeal of the irrational also seems to grow stronger in uncertain times and this may prove far more dangerous for human survival than the currently reviled arrogance of rationalist approaches to solving human problems. Finally, because material conditions can change unpredictably, knowledge of what has happened in the past does not provide certain guidance for the future, and never less so than in times of rapid technological and social change. None of these considerations bodes well for the future. Yet what we know about our collective past suggests that co-operation, generosity, and planning have been as important in shaping our development as have competition and rivalry. To what extent as a species can we cope with ever faster change and ever more powerful technologies? And to what extent can we go on defying the second law of thermodynamics by evolving ever more elaborate and energy-expensive lifestyles? Despite these imponderables, it seems possible to suggest on the basis of past experience that, when it comes to survival, planning our collective future will prove far more efficacious than entrusting it to metaphysical entities such as the invisible hand. The key qualities we must learn to cultivate are foresight, justice, and self-restraint. These attributes are ones that have developed as a result of human experiences during the course of our biological and sociocultural evolution.

12
Conclusion

The idea that the evolution of ever more complex societies and cultures is a key feature of human history has been a major concept animating Western social, political, and economic thought since the eighteenth century. As an important component of Enlightenment philosophy, antedating biological evolution, the doctrine of sociocultural progress helped to direct the revolutionary transformations that gave rise to modern societies based on the ideals of all humans sharing the same freedoms, responsibilities, and opportunities for self-realization. Almost from the beginning, however, evolutionist thought was assailed by critics who maintained that its universalistic tendencies devalued cultural diversity and the traditional associations that pro de 1 ch of the meaning to everyday life. Today sociocultural evolution is being attacked more ferociously than ever before for allegedly being inherently ethnocentric, racist, and factually untenable.

It is generally agreed that, contrary to what was believed in the eighteenth and nineteenth centuries, no tendency towards increasing sociocultural complexity is inherent in either the cosmic order or human nature. Nor does increasing sociocultural complexity automatically result in general progress, in the sense of improvement in all aspects of social life, or benefit everyone in a society equally. On the contrary, technological progress has produced much individual suffering and resulted in the oppression of whole peoples by their more advanced neighbours. Yet, where technological progress has occurred, it has been accompanied by the development of ever larger and more internally differentiated societies. Within such societies nature is conceptualized in terms that are increasingly different from those used to conceptualize human society, and there is an increasingly detailed and hopefully more realistic understanding of human behaviour. Living successfully with more complex technologies also requires increasing individual self-discipline and self-restraint. Technological progress has sustained an extraordinary increase in human numbers and has enabled human beings – at least in the short run – to dominate and exploit the world's ecosystem.

Human behaviour is unique in the extent to which it is learned and symbolically mediated. This makes it unusually flexible and subject to modification on the basis of indirect, linguistically transmitted experience as well as direct personal experience. While culture serves to adapt human populations to their natural and cultural environments, the self-reflexivity of symbolic thought makes this adaptation extraordinarily complex as well as relatively easily modified. There is no evidence that human beings are always seeking easier and more secure lives by learning to control nature more effectively, as many nineteenth-century evolutionists believed. There is also no evidence that human beings resist change unless it is forced on them by environmental challenges, as many neo-evolutionists argued. Most human behaviour is guided by concepts that are transmitted from one generation to another and modified as ecological circumstances change or opportunities arise for certain individuals or groups to profit from change at the expense of their neighbours. Technological change takes place within the context of specific societies and modifies existing beliefs and knowledge in ways that are sometimes predictable and sometimes highly idiosyncratic. This cultural equivalent of descent with modification may explain the development of cultural diversity. It does not, however, explain the linear tendencies that prevail in human history to a much greater extent than in biological evolution.

A particular institution may develop in one society that functions more effectively than does an analogous institution in another society. So long as these societies are not in contact, this difference is of no consequence. Yet, when societies begin to compete with one another, the society that has more efficient institutions will be favoured. In biological evolution the species that is better adapted to a particular environment expands at the expense of one that is less fit. People living in less competitively advantaged societies may try to alter them to resemble more closely a more successful competitor. This process appears to be made easier by a common tendency for individuals to copy what they perceive to be more successful behaviour in others. Alternatively, and this happens most commonly when the technological and organizational gap between societies is great, the more complex society may come to dominate and either control or completely absorb weaker ones. In both instances, cross-cultural variation is reduced or eliminated. This is the aspect of sociocultural evolution that most resembles the operation of natural selection in biological evolution. The capacity of people to cope with difficulties by inventing new forms of behaviour or imitating those of more successful competitors, as well as the capacity of more powerful societies to dominate and reshape weaker ones, ensure that there is considerably more linearity to sociocultural evolution than there is to biological evolution.

While change occurred slowly and the capacity of neighbouring societies to transform one another was limited as long as all human societies remained at a relatively simple hunter–gatherer level, the last 10,000 years have witnessed ever-accelerating social and cultural change and the ability of more complex societies to bring about sociocultural transformations over ever larger areas, until today every society in the world is part of a single interactive economic and communications network. Within this network, smaller, less 'evolved' societies have been influenced to ever greater degrees by technologically more powerful and 'evolved' ones. The past few decades have seen the last surviving hunter–gatherer societies drawn into a world economy. Linearity has not always been a key feature of human history, but over time it has become more pronounced.

The realization that this is so provides an explanation of much past human behaviour and a framework for understanding the colonization and exploitation of people who are for historical and cultural reasons less well equipped to defend their own interests both within individual societies and among interacting ones. Such an understanding does not condone exploitation on the grounds that it is inevitable or natural. On the contrary, an understanding of evolutionary processes may indicate the need to curb such behaviour in order to create a world system that can remain viable. The increasing rate of change and the ever-greater disruptive impacts that such change is having on the social order and the ecosystem suggest that humanity has reached a point where its future can no longer be left to chance. In the past, serious errors of judgement never threatened the future of the entire human species or of the whole ecosystem; today they do. This suggests that more emphasis has to be placed on planning and informed decision-making at all levels, as well as on creating a global society in which planning is both possible and effective at local and regional as well as national and global levels. While this idea runs counter to the prevailing neo-conservative ideology, it stresses a capacity for foresight, calculation, and decision-making that is unique to human beings and largely responsible for their rapid numerical expansion and current domination, for better or worse, of the terrestrial ecosystem.

Whether we regard greater complexity as good or bad, it has been an increasingly prominent feature of sociocultural evolution for the past several millennia. The ultimate purpose of a sociocultural evolutionary perspective is not to justify a specific course of action on the grounds that it is inevitable (a proposition that is scientifically untenable in so far as statements about the future are unverifiable until that future has been realized). As a guide to the future, an evolutionary approach should encourage a scientific understanding of sociocultural change, combined with an ethical commitment to formulating policies that will enrich both the lives of individuals and the societies in which all human beings live. Morality and self-control are qualities that must continue to evolve in order to ensure human survival. Whether human beings can acquire sufficient foresight, self-discipline, and social restraint to protect the environment, and avoid social chaos as a result of growing immiseration of human beings, will determine not only the kind of future humanity as a whole will have, but whether or not they will have a future at all.

An evolutionary approach should consider the consequences of alternative courses of action in the hope that human beings may choose the course that best assures their collective survival and well-being. In that way, growing complexity may result in true progress, which involves producing the greatest possible good for the greatest number, without inflicting injustice on anyone. The belief that this is possible must not be lost sight of amidst the seductive temptations of a relativism that, despite its valid points, can all too easily descend into amorality and defeatism and end up being used to justify greed and social irresponsibility. Progress is not inherent in sociocultural evolution, but something that individuals must plan and be prepared to fight for if it is to become a reality. If our own and future generations are to enjoy a life that is truly worth living, the belief that we can transcend current limitations and build a more just, secure, and satisfying future must remain a guiding principle and our most important inheritance from the Enlightenment.

Bibliographic Note

This note is a guide for those who wish to read more about the development of thought concerning sociocultural evolution. Despite the importance of this topic, the number of works devoted exclusively to it is not as large as might be expected.

The most general history of sociocultural evolution remains Marvin Harris's *The Rise of Anthropological Theory* (1968). Although written three decades ago from a neo-evolutionary and largely anthropological point of view, this book provides the best general survey of the history of sociocultural evolution and of the alternative theories with which evolutionists have had to contend. Another, more narrowly focused, general survey is Stephen Sanderson's *Social Evolutionism: A Critical History* (1990), which compares and contrasts a broad range of sociocultural evolutionary theories from the early nineteenth century to the present. Sanderson presents detailed summaries of different approaches to sociocultural evolution and of the ideas of individual evolutionists. His book also contains an excellent bibliography.

There is no definitive study of sociocultural evolution prior to the nineteenth century. Stephen Toulmin and June Goodfield provide an excellent account of the growing awareness of evolutionary change in the natural and social realms in their book *The Discovery of Time* (1966). Paolo Rossi deals with the beginnings of this awareness in *The Dark Abyss of Time* (1984). More specifically focused coverage of early sociocultural evolution can be found in

Margaret Hodgen's *Early Anthropology in the Sixteenth and Seventeenth Centuries* (1964) and James Slotkin's *Readings in Early Anthropology* (1965). Many works deal with the Enlightenment. Among these we suggest Norman Hampson, *The Enlightenment* (1982) and Ulrich Im Hof, *The Enlightenment* (1994). The specialized development of Scottish Enlightenment thinking is described by Gladys Bryson in *Man and Society: The Scottish Inquiry of the Eighteenth Century* (1945). The origins of German romanticism are traced in Frederick Barnard's *Herder's Social and Political Thought* (1965) and Frederick Beiser's *Enlightenment, Revolution, and Romanticism* (1992).

Much more has been written about nineteenth-century sociocultural evolutionism. John Burrow's *Evolution and Society* (1966) offers a general account of Victorian social evolution and of the hold that this idea had on the minds of Victorian intellectuals. Peter Bowler's *The Invention of Progress* (1989) deals with the conflict between continuous and cyclical models of sociocultural progress and how they were influenced by findings in geology and biology. Developments in geology and biology are analysed in greater detail in Bowler's *Evolution: The History of an Idea* (1984) and *Theories of Human Evolution* (1986). Adrian Desmond studies the political implications of the concept of evolution in early nineteenth-century Britain in *The Politics of Evolution* (1989). George Stocking's *Victorian Anthropology* (1987) analyses how British experiences of cultural change during the industrial revolution shaped their evolutionary views.

In *Evolution and Social Life* (1986) Tim Ingold traces in detail how the concept of sociocultural evolution has been handled theoretically from the early nineteenth century to the present. In particular, he examines the impact of Darwinism and of the more recent formulation of a fundamental opposition between social process and cultural form. Although it is not a historical study, the history of the latter dichotomy is also addressed in Christopher Hallpike's *The Principles of Social Evolution* (1986). Maurice Bloch deals with the Marxist origins of this dichotomy in his book *Marxism and Anthropology* (1983). Elman Service's *A Century of Controversy* (1985) examines, from the viewpoint of a leading neo-evolutionist, the dominant ideas that have influenced the interpretation of sociocultural change since the 1860s. In *Race, Culture, and Evolution* (1968), George Stocking traces the impact of racism on sociocultural evolution and alternative explanations of cultural development in the late nineteenth and early twentieth centuries. Much of the sociocultural evolutionary literature of the twentieth century that relates to the structured properties of cultural trait acquisition, transmission, and modification is surveyed in Robert Boyd and Peter Richerson's *Culture and the Evolutionary Process* (1985). They provide an unpolemical but highly critical assessment of the sociobiological approach to sociocultural evolution.

The modern rejection of sociocultural evolution has not yet produced a major critical work, but current criticisms of sociocultural evolutionism in general or of neo-evolutionism in particular can be found in Ernest Gellner's *Plough, Sword and Book* (1988), Anthony Giddens's *The Constitution of Society* (1984), Michael Mann's *The Sources of Social Power* (1986), and many of the papers in Norman Yoffee and Andrew Sherratt's *Archaeological Theory* (1993).

References

Adams, Richard N. (1988), *The Eighth Day: Social Evolution as the Self-Organization of Energy*. Austin, TX: University of Texas Press.

Adams, Robert McC. (1965), *Land behind Baghdad: A History of Settlement on the Diyala Plains*. Chicago, IL: University of Chicago Press.

—— (1974), Anthropological perspectives on ancient trade. *Current Anthropology*, 15, pp. 239–58.

—— (1996), *Paths of Fire: An Anthropologist's Inquiry into Western Technology*. Princeton, NJ: Princeton University Press.

Allen, Arthur (1992), Ancient Mayan rituals, beliefs are alive today, scientists say. *The Gazette* (Montreal), 23 May.

Anderson, Benedict (1983), *Imagined Communities: Reflections on the Origin and Spread of Nationalism*. London: Verso.

Ardrey, Robert (1961), *African Genesis: A Personal Investigation into the Animal Origins and Nature of Man*. New York: Atheneum.

—— (1966), *The Territorial Imperative: A Personal Inquiry into the Animal Origins of Property and Nations*. New York: Atheneum.

Aronow, Saul (1963), *The Fallen Sky: Medical Consequences of Thermonuclear War*. New York: Hill and Wang.

Ashworth, William (1986), *The Late, Great Lakes: An Environmental History*. New York: Knopf.

Bachofen, Johann (1861), *Das Mutterrecht: Eine Untersuchung über die Gynaikokratie der alten Welt nach ihrer religiösen und rechtlichen Natur*. Stuttgart: Krais und Hoffmann.

Bagehot, Walter (1872), *Physics and Politics: Or, Thoughts on the Applica-

tion of the Principles of 'Natural Selection' and 'Inheritance' to Political Society. London: H. S. King.

Bailey, Anne M., and Josep R. Llobera (1981), *The Asiatic Mode of Production: Science and Politics.* London: Routledge and Kegan Paul.

Barnard, Frederick M. (1965), *Herder's Social and Political Thought: From Enlightenment to Nationalism.* Oxford: Oxford University Press.

Barnes, Barry (1974), *Scientific Knowledge and Sociological Theory.* London: Routledge and Kegan Paul.

—— (1977), *Interests and the Growth of Knowledge.* London: Routledge and Kegan Paul.

Baum, Gregory (1996), *Karl Polanyi on Ethics and Economics.* Montreal: McGill-Queen's University Press.

Beckerman, Wilfred (1995), *Small is Stupid: Blowing the Whistle on the Greens.* London: Duckworth.

Beiser, Frederick C. (1992), *Enlightenment, Revolution, and Romanticism: The Genesis of Modern German Political Thought, 1790–1800.* Cambridge, MA: Harvard University Press.

Bellini, James (1986), *High Tech Holocaust.* Newton Abbot: David and Charles.

Benedict, Ruth (1934), *Patterns of Culture.* Boston, MA: Houghton Mifflin.

Berger, Carl (1970), *The Sense of Power: Studies in the Ideas of Canadian Imperialism, 1867–1914.* Toronto: University of Toronto Press.

Bieder, Robert E. (1986), *Science Encounters the Indian, 1820–1880: The Early Years of American Ethnology.* Norman, OK: University of Oklahoma Press.

Binford, Lewis L. (1962), Archaeology as anthropology. *American Antiquity,* 28, pp. 217–25.

—— (1972), *An Archaeological Perspective.* New York: Seminar Press.

—— (ed.) (1977), *For Theory Building in Archaeology: Essays on Faunal Remains, Aquatic Resources, Spatial Analysis and Systemic Modelling.* New York: Academic Press.

—— (1981), *Bones: Ancient Men and Modern Myths.* New York: Academic Press.

—— (1983), *In Pursuit of the Past: Decoding the Archaeological Record.* London: Thames and Hudson.

Blanton, Richard E., Stephen A. Kowalewski, Gary M. Feinman, and Jill Appel (1981), *Ancient Mesoamerica: A Comparison of Change in Three Regions.* Cambridge: Cambridge University Press.

Blanton, Richard E., Gary M. Feinman, Stephen A. Kowalewski, and

Peter N. Peregrine (1996), A dual-processual theory for the evolution of Mesoamerican civilization. *Current Anthropology*, 37, pp. 1–14.

Bloch, Maurice (1983), *Marxism and Anthropology: The History of a Relationship*. Oxford: Oxford University Press.

Bloom, Howard (1995), The global warming scare is overblown. *The Gazette* (Montreal), 11 November, p. B-6.

Boserup, Ester (1965), *The Conditions of Agricultural Growth: The Economics of Agrarian Change under Population Pressure*. Chicago, IL: Aldine.

—— (1981), *Population and Technological Change: A Study of Long-Term Trends*. Chicago, IL: University of Chicago Press.

Bowler, Peter J. (1984), *Evolution: The History of an Idea*. Berkeley, CA: University of California Press.

—— (1986), *Theories of Human Evolution: A Century of Debate, 1844–1944*. Baltimore, MD: Johns Hopkins University Press.

—— (1989), *The Invention of Progress: The Victorians and the Past*. Oxford: Basil Blackwell.

—— (1992), From 'savage' to 'primitive': Victorian evolutionism and the interpretation of marginalized peoples. *Antiquity*, 66, pp. 721–29.

Boyd, Robert, and Peter J. Richerson (1985), *Culture and the Evolutionary Process*. Chicago, IL: University of Chicago Press.

Braidwood, Robert J. (1974), The Iraq Jarmo Project. In Gordon R. Willey (ed.), *Archaeological Researches in Retrospect*, pp. 59–83. Cambridge, MA: Winthrop.

Branwell, Anna (1989), *Ecology in the 20th Century: A History*. New Haven, CT: Yale University Press.

—— (1994), *The Fading of the Greens: The Decline of Environmental Politics in the West*. New Haven, CT: Yale University Press.

Brenner, Reuven (1994), *Labyrinths of Prosperity: Economic Follies, Democratic Remedies*. Ann Arbor, MI: University of Michigan Press.

Brohman, John (1996), *Popular Development: Rethinking the Theory and Practice of Development*. Oxford: Basil Blackwell.

Bronson, Bennet (1972), Farm labor and the evolution of food production. In Brian Spooner (ed.), *Population Growth: Anthropological Implications*, pp. 190–218. Cambridge, MA: MIT Press.

Brose, David S. (1993), Changing paradigms in the explanation of Southeastern prehistory. In Jay K. Johnson (ed.), *The Development of Southeastern Archaeology*, pp. 1–17. Tuscaloosa, AL: The University of Alabama Press.

Brown, Donald E. (1991), *Human Universals*. Philadelphia, PA: Temple University Press.

Brown, Lester R., and Edward C. Wolf (1984), *Soil Erosion: Quiet Crisis in the World Economy*. Washington, DC: Worldwatch Institute.

Bryson, Gladys (1945), *Man and Society: The Scottish Inquiry of the Eighteenth Century*. Princeton, NJ: Princeton University Press.

Buckle, Henry T. (1857), *History of Civilization in England*, vol. 1. London: J. W. Parker.

Bunge, Mario A. (1996), *Finding Philosophy in Social Science*. New Haven, CT: Yale University Press.

Burger, Richard, and Lucy Salazar-Burger (1993), The place of dual organization in early Andean ceremonialism: A comparative review. In Luis Millones and Yoshio Onuki (eds), *El Mundo Ceremonial Andino*, pp. 97–116. Osaka: National Museum of Ethnology, Senri Ethnological Studies, 37.

Burrow, John W. (1966), *Evolution and Society: A Study in Victorian Social Theory*. Cambridge: Cambridge University Press.

Carneiro, Robert L. (1970a), Scale analysis, evolutionary sequences, and the rating of cultures. In Raoul Naroll and Ronald Cohen (eds), *A Handbook of Method in Cultural Anthropology*, pp. 834–71. New York: Columbia University Press.

—— (1970b), A theory of the origin of the state. *Science*, 169, pp. 733–8.

—— (1973), The four faces of evolution. In John J. Honigmann (ed.), *Handbook of Social and Cultural Anthropology*, pp. 89–110. Chicago, IL: Rand-McNally.

Carrithers, Michael (1992), *Why Humans Have Cultures: Explaining Anthropology and Social Diversity*. Oxford: Oxford University Press.

Carson, Rachel (1962), *Silent Spring*. Boston, MA: Houghton Mifflin.

[Chambers, Robert] (1844), *Vestiges of the Natural History of Creation*. London: John Churchill.

Chambers, Robert (1983), *Rural Development: Putting the Last First*. London: Longman.

Childe, V. Gordon (1928), *The Most Ancient East: The Oriental Prelude to European Prehistory*. London: Kegan Paul.

—— (1930), *The Bronze Age*. Cambridge: Cambridge University Press.

—— (1936), *Man Makes Himself*. London: Watts (pages cited from 4th edn, 1965).

—— (1939), *The Dawn of European Civilization*, 3rd edn. London: Kegan Paul.

—— (1942), *What Happened in History*. Harmondsworth: Penguin.

—— (1946), *Scotland before the Scots*. London: Methuen.

—— (1947), *History*. London: Cobbett.

—— (1949), *Social Worlds of Knowledge*. London: Oxford University Press.

—— (1950), *Magic, Craftsmanship and Science*. Liverpool: Liverpool University Press.

—— (1951), *Social Evolution*. New York: Schuman.

—— (1956), *Society and Knowledge: The Growth of Human Traditions*. New York: Harper.

—— (1958), *The Prehistory of European Society*. Harmondsworth: Penguin.

Clark, Grahame (1986), *Symbols of Excellence: Precious Materials as Expressions of Status*. Cambridge: Cambridge University Press.

Clements, Robert J. (1974), Michelangelo. In *Encyclopaedia Britannica*, 15th edn, Macropaedia, 12, pp. 97–102.

Cohen, Mark N. (1977), *The Food Crisis in Prehistory: Overpopulation and the Origins of Agriculture*. New Haven, CT: Yale University Press.

Condorcet, Marie-Jean de Caritat, Marquis de (1795), *Esquisse d'un tableau historique des progrès de l'esprit humain*. Paris: Agasse.

Cordell, Linda S., and Fred Plog (1979), Escaping the confines of normative thought: A reevaluation of Puebloan prehistory. *American Antiquity*, 44, pp. 405–29.

Cowgill, George L. (1975), On causes and consequences of ancient and modern population changes. *American Anthropologist*, 77, pp. 505–25.

Crawford, Robert (1992), *Devolving English Literaure*. New York: Oxford University Press.

Däniken. *See* von Däniken.

Dart, Raymond A. (1949), The predatory implemental technique of *Australopithecus*. *American Journal of Physical Anthropology*, 7, pp. 1–38.

Darwin, Charles G. (1938), Logic and probability in physics. *Report of the Annual Meeting of the British Association for the Advancement of Science for 1938*, pp. 21–34.

Darwin, Charles R. [1839] (1959), *The Voyage of the Beagle*, Introd. by H. Graham Cannon. Everyman's Library, 104. London: Dent.

—— (1859), *On the Origin of Species by Means of Natural Selection: Or, the Preservation of Favoured Races in the Struggle for Life*. London: John Murray.

—— (1874), *The Descent of Man, and Selection in Relation to Sex*, 2nd edn. London: John Murray.

Dawson, John William (1888), *Fossil Men and their Modern Representatives: An Attempt to Illustrate the Characters and Condition of Pre-historic Men in Europe, by those of the American Races*, 3rd edn. Montreal: Dawson.

Deneven, William M. (1992), The pristine myth: The landscape of the Americas in 1492. *Annals of the Association of American Geographers*, 82, pp. 369–85.

Desmond, Adrian J. (1989), *The Politics of Evolution: Morphology, Medicine, and Reform in Radical London*. Chicago, IL: University of Chicago Press.

—— (1994), *Huxley: The Devil's Disciple*. London: Michael Joseph.

Desmond, Adrian, and James Moore (1992), *Darwin*. Harmondsworth: Penguin.

Devall, Bill, and George Sessions (1985), *Deep Ecology*. Salt Lake City, UT: G. M. Smith.

Diamond, Stanley (1974), *In Search of the Primitive: A Critique of Civilization*. New Brunswick, NJ: Transaction Books.

Dietler, Michael (1994), 'Our ancestors the Gauls': Archaeology, ethnic nationalism, and the manipulation of Celtic identity in modern Europe. *American Anthropologist*, 96, pp. 584–605.

Disraeli, Benjamin (1845), *Sybil; or, The Two Nations*. London: H. Colburn.

Dobyns, Henry F. (1983), *Their Number Become Thinned: Native American Population Dynamics in Eastern North America*. Knoxville, TN: University of Tennessee Press.

Dotto, Lydia (1986), *Planet Earth in Jeopardy: Environmental Consequences of Nuclear War*. New York: John Wiley.

Drengson, Alan, and Yuichi Inoue (eds) (1995), *The Deep Ecology Movement: An Introductory Anthology*. Berkeley, CA: North Atlantic Books.

Driver, Harold E. (1966), Geographical-historical *versus* psychofunctional explanations of kin avoidances. *Current Anthropology*, 7, pp. 131–82.

Dumont, Louis (1994), *German Ideology: From France to Germany and Back*. Chicago, IL: University of Chicago Press.

Dunn, John (1993), *Western Political Theory in the Face of the Future*, 2nd edn. Cambridge: Cambridge University Press.

Dunn, Stephen P. (1982), *The Fall and Rise of the Asiatic Mode of Production*. London: Routledge and Kegan Paul.

Dunnell, Robert C. (1980), Evolutionary theory and archaeology. In

Michael B. Schiffer (ed.), *Advances in Archaeological Method and Theory*, 3, pp. 35–99. New York: Academic Press.

Durkheim, Emile (1893), *De la division du travail social*. Paris: Alcan.

—— (1895), *Les règles de la méthode sociologique*. Paris: Alcan.

Easterbrook, Gregg (1995), *A Moment on the Earth: The Coming Age of Environmental Optimism*. New York: Viking.

Ehrlich, Paul R. (1968), *The Population Bomb*. New York: Ballantine.

Ehrlich, Paul R., and Anne Erhlich (1981), *Extinction: The Causes and Consequences of the Disappearance of Species*. New York: Random House.

Eiseley, Loren C. (1958), *Darwin's Century and the Men Who Discovered It*. Garden City, NY: Doubleday.

Eisenstadt, Shmuel N. (ed.) (1986), *The Origins and Diversity of Axial Age Civilizations*. Albany, NY: State University of New York Press.

Elias, Norbert (1978), *The Civilizing Process*, vol. 1, *The Development of Manners: Changes in the Code of Conduct and Feeling in Early Modern Times*. New York: Urizen Books.

—— (1982), *The Civilizing Process*, vol. 2, *State Formation and Civilization*. Oxford: Basil Blackwell.

Elliott, Larry (1996), Bridging the North–South divide: Global equality is central to the next phase of industrial revolution. *Guardian Weekly*, 155 (6), 11 August, p. 14.

Engels, Frederick [1844] (1973), *The Condition of the Working-Class in England, From Personal Observation and Authentic Sources*. Moscow: Progress Publishers.

—— [1884] (1972), *The Origin of the Family, Private Property and the State in the Light of the Researches of Lewis Henry Morgan*, ed. Eleanor B. Leacock. New York: International Publishers.

Ensminger, Jean, and Jack Knight (1997), Common property, bride-wealth, and clan exogamy. *Current Anthropology*, 38, pp. 1–24.

Evans, Arthur J. (1896), The 'Eastern Question' in anthropology. *Proceedings of the British Association for the Advancement of Science for 1896*, pp. 906–22.

Evans-Pritchard, Edward E. (1949), *The Sanusi of Cyrenaica*. Oxford: Oxford University Press.

—— (1962), Anthropology and history. In E. E. Evans-Pritchard, *Essays in Social Anthropology*, pp. 45–65. London: Faber.

Fairchild, Hoxie N. (1928), *The Noble Savage: A Study in Romantic Naturalism*. New York: Columbia University Press.

Fairservis, Walter A. (1959), *The Origins of Oriental Civilization*. New York: New American Library.

Fash, William L. (1991), *Scribes, Warriors and Kings: The City of Copán and the Ancient Maya*. London: Thames and Hudson.

Feit, Harvey A. (1978), Waswanipi Realities and Adaptations: Resource Management and Cognitive Structure. PhD thesis, McGill University, Montreal.

Feyerabend, Paul K. (1975), *Against Method: Outline of an Anarchistic Theory of Knowledge*. London: NLB.

Flannery, Kent V. (1972), The cultural evolution of civilizations. *Annual Review of Ecology and Systematics*, 3, pp. 399–426.

—— (1995), Prehistoric social evolution. In Carol R. Ember and Melvin Ember (eds), *Research Frontiers in Anthropology*, pp. 1–26. Englewood Cliffs, NJ: Prentice-Hall.

Flannery, Kent V., and Joyce Marcus (1976), Formative Oaxaca and the Zapotec cosmos. *American Scientist*, 64, pp. 374–83.

—— (eds) (1983), *The Cloud People: Divergent Evolution of the Zapotec and Mixtec Civilizations*. New York: Academic Press.

Ford, Tom (1996), Multinationals won't give up easily. *Beacon Herald* (Stratford, Ont.), 7 July.

Forge, Anthony (1972), Normative factors in the settlement size of Neolithic cultivators (New Guinea). In Peter J. Ucko, Ruth Tringham, and G. W. Dimbleby (eds), *Man, Settlement and Urbanism*, pp. 363–76. London: Duckworth.

Francis, Daniel, and Toby Morantz (1983), *Partners in Furs: A History of the Fur Trade in Eastern James Bay, 1600–1870*. Montreal: McGill-Queen's University Press.

Frankfort, Henri (1948), *Kingship and the Gods: A Study of Ancient Near Eastern Religion as the Integration of Society and Nature*. Chicago, IL: University of Chicago Press.

—— (1956), *The Birth of Civilization in the Near East*. Garden City, NY: Doubleday.

Frankfort, Henri, H. A. Frankfort, John A. Wilson, and Thorkild Jacobsen (1949), *Before Philosophy: The Intellectual Adventure of Ancient Man*. Harmondsworth: Penguin.

Franklin, Ursula M. (1990), *The Real World of Technology*. Toronto: CBC Publications.

Frazer, James G. (1911–15), *The Golden Bough: A Study in Magic and Religion*, 8 vols, 3rd edn. New York: Macmillan.

Freeman, Edward A. (1867–79), *The History of the Norman Conquest of England: Its Causes and its Results*, 6 vols. Oxford: Clarendon Press.

Freeman, Milton M. R. (1974), *People Pollution: Sociologic and Ecologic Viewpoints on the Prevalence of People*. Montreal: McGill-Queen's University Press.

Freidel, David, Linda Schele, and Jay Parker (1993), *Maya Cosmos: Three Thousand Years on the Shaman's Path*. New York: William Morrow.

Frick, Wilhelm (1934), The teaching of history and prehistory in Germany. *Nature*, 133, pp. 298–9.

Fried, Morton H. (1967), *The Evolution of Political Society: An Essay in Political Anthropology*. New York: Random House.

—— (1975), *The Notion of Tribe*. Menlo Park, CA: Cummings.

Friedman, Milton (1962), *Capitalism and Freedom*. Chicago, IL: University of Chicago Press.

Fukuyama, Francis (1992), *The End of History and the Last Man*. New York: Free Press.

Fyfe, William S. (1985), Global change: What should Canada do? *Transactions of the Royal Society of Canada*, 23, pp. 193–9.

Galton, Francis (1869), *Hereditary Genius: An Inquiry into Its Laws and Consequences*. London: Macmillan.

Geertz, Clifford (1973), *The Interpretation of Cultures: Selected Essays*. New York: Basic Books.

Gellner, Ernest (1983), *Nations and Nationalism*. Oxford: Basil Blackwell.

—— (1988), *Plough, Sword and Book: The Structure of Human History*. London: Collins Harvill.

Gibbon, Edward (1776–88), *The History of the Decline and Fall of the Roman Empire*, 6 vols. London: W. Strahan and T. Cadell.

Giddens, Anthony (1981), *A Contemporary Critique of Historical Materialism*. London: Macmillan.

—— (1984), *The Constitution of Society: Outline of the Theory of Structuration*. Berkeley, CA: University of California Press.

Gobineau, Joseph-Arthur, Comte de (1853–5), *Essai sur l'inégalité des races humaines*, 4 vols. Paris: Didot.

Godelier, Maurice (1977), *Perspectives in Marxist Anthropology*. Cambridge: Cambridge University Press.

—— (1978), Economy and religion: An evolutionary optical illusion. In Jonathan Friedman and Michael J. Rowlands (eds), *The Evolution of Social Systems*, pp. 3–11. London: Duckworth.

Goffman, Erving (1963), *Behavior in Public Places: Notes on the Social Organization of Gatherings*. New York: Free Press.

Goguet, Antoine-Ives (1761), *The Origin of Laws, Arts, and Sciences, and their Progress among the Most Ancient Nations*, 3 vols. Edinburgh: A. Donaldson and J. Reid.

Goldenweiser, Alexander A. (1913), The principle of limited possibilities in the development of culture. *Journal of American Folk-Lore*, 26, pp. 259–90.

Golson, Jack, and D. S. Gardner (1990), Agriculture and sociopolitical organization in New Guinea highlands prehistory. *Annual Review of Anthropology*, 19, pp. 395–417.

Goodenough, Ward H. (1972), Social implications of population control. *Expedition*, 14(3), pp. 11–14.

Gould, Stephen J. (1996), *Full House: The Spread of Excellence from Plato to Darwin*. New York: Harmony Books.

Grant, George P. (1965), *Lament for a Nation: The Defeat of Canadian Nationalism*. Toronto: McClelland and Stewart.

Grant, Madison (1916), *The Passing of the Great Race; Or, The Racial Basis of European History*. New York: Scribner's.

Gräslund, Bo (1987), *The Birth of Prehistoric Chronology: Dating Methods and Dating Systems in Nineteenth-Century Scandinavian Archaeology*. Cambridge: Cambridge University Press.

Grayson, Donald K. (1983), *The Establishment of Human Antiquity*. New York: Academic Press.

Green, Michael D. (1996), The expansion of European colonization to the Mississippi Valley, 1780–1880. In Bruce G. Trigger and Wilcomb E. Washburn (eds), *The Cambridge History of the Native Peoples of the Americas*, vol. 1, *North America*, Part 1, pp. 461–538.

Habermas, Jürgen (1971), *Knowledge and Human Interests*. Boston, MA: Beacon Press.

—— (1979), *Communication and the Evolution of Society*. Boston, MA: Beacon Press.

—— (1984), *The Theory of Communicative Action*, vol. 1. Boston, MA: Beacon Press.

Hägerstrand, Torsten (1967), *Innovation Diffusion as a Spatial Process*. Chicago, IL: University of Chicago Press.

Hall, John A. (1986), *Powers and Liberties: The Causes and Consequences of the Rise of the West*. Harmondsworth: Penguin.

Hallpike, Christopher R. (1979), *The Foundations of Primitive Thought*. Oxford: Oxford University Press.

—— (1986), *The Principles of Social Evolution*. Oxford: Oxford University Press.

Hamill, James F. (1990), *Ethno-logic: The Anthropology of Human Reasoning*. Urbana, IL: University of Illinois Press.

Hampson, Norman (1982), *The Enlightenment*. Harmondsworth: Penguin.

Hardin, Garrett J. (1968), The tragedy of the commons. *Science*, 162, pp. 1243–8.

Hariot, Thomas [1588] (1951), *A Brief and True Report of the New Found Land of Virginia*, facsimile edn. Ann Arbor, MI: Clements Library Associates.

Harris, Marvin (1968), *The Rise of Anthropological Theory: A History of Theories of Culture*. New York: Crowell.

—— (1974), *Cows, Pigs, Wars and Witches: The Riddles of Culture*. New York: Random House.

—— (1977), *Cannibals and Kings: The Origins of Cultures*. New York: Random House.

—— (1979), *Cultural Materialism: The Struggle for a Science of Culture*. New York: Random House.

Hart, Hornell (1959), Social theory and social change. In Llewellyn Gross (ed.), *Symposium on Sociological Theory*, pp. 196–238. Evanston, IL: Row, Peterson.

Harvey, David (1990), *The Condition of Postmodernity: An Enquiry into the Origins of Cultural Change*. Oxford: Basil Blackwell.

Harvey, Robert (1995), *The Return of the Strong: Drift to Global Disorder*. London: Macmillan.

Hassig, Ross (1985), *Trade, Tribute, and Transportation: The Sixteenth-Century Political Economy of the Valley of Mexico*. Norman, OK: University of Oklahoma Press.

Hatch, Elvin (1983), *Culture and Morality: The Relativity of Values in Anthropology*. New York: Columbia University Press.

Haven, Samuel (1856), *Archaeology of the United States*. Washington, DC: Smithsonian Contributions to Knowledge, 8(2).

Hayek, Friedrich A. von (1960), *The Constitution of Liberty*. Chicago, IL: University of Chicago Press.

Heizer, Robert F. (ed.) (1962), *Man's Discovery of his Past: Literary Landmarks in Archaeology*. Englewood Cliffs, NJ: Prentice-Hall.

Herold, J. Christopher (1962), *Bonaparte in Egypt*. New York: Harper and Row.

Herskovits, Melville J. (1953), *Franz Boas: The Science of Man in the Making*. New York: Scribner.

Hobbes, Thomas [1651] (1955), *Leviathan, Or the Matter, Forme and Power of a Commonwealth Ecclesiasticall and Civil*, ed. Michael Oakeshott. Oxford: Basil Blackwell.

Hobhouse, Leonard T., Gerald C. Wheeler, and Morris Ginsberg (1915), *The Material Culture and Social Institutions of the Simpler Peoples: An Essay in Correlation*. London: Chapman and Hall.

Hobsbawm, Eric J. (ed.) (1964), *Karl Marx: Pre-Capitalist Economic Formations*. London: Lawrence and Wishart.

—— (1994), *The Age of Extremes: A History of the World, 1914–1991*. New York: Pantheon Books.

Hodder, Ian (1982), *Symbols in Action: Ethnoarchaeological Studies of Material Culture*. Cambridge: Cambridge University Press.

—— (1991), *Reading the Past: Current Approaches to Interpretation in Archaeology*, 2nd edn. Cambridge: Cambridge University Press.

Hodgen, Margaret T. (1964), *Early Anthropology in the Sixteenth and Seventeenth Centuries*. Philadelphia, PA: University of Pennsylvania Press.

Horsman, Mathew, and Andrew Marshall (1994), *After the Nation-State: Citizens, Tribalism, and the New World Disorder*. London: Harper Collins.

Hosler, Dorothy, Jeremy A. Sabloff, and Dale Runge (1977), Simulation model development: A case study of the Classic Maya collapse. In Norman Hammond (ed.), *Social Process in Maya History*, pp. 553–90. New York: Academic Press.

Hughes, J. Donald (1975), *Ecology in Ancient Civilizations*. Albuquerque, NM: University of New Mexico Press.

Hunt, Lynn (1989), Introduction: History, culture, and text. In Lynn Hunt (ed.), *The New Cultural History*, pp. 1–22. Berkeley, CA: University of California Press.

Huxley, Julian (1960), *Knowledge, Morality, and Destiny* (original title *New Bottles for New Wine*). New York: New American Library.

Huxley, Thomas H. [1893] (1947), Evolution and ethics. In Thomas H. Huxley and Julian Huxley, *Evolution and Ethics, 1893–1943*, pp. 60–102. London: Pilot Press.

Ikawa-Smith, Fumiko (ed.) (1978), *Early Paleolithic in South and East Asia*. The Hague: Mouton.

Im Hof, Ulrich (1994), *The Enlightenment*. Oxford: Basil Blackwell.

Ingold, Tim (1986), *Evolution and Social Life*. Cambridge: Cambridge University Press.

Innis, Harold A. (1951), *The Bias of Communication*. Toronto: University of Toronto Press.

Irvine, William (1955), *Apes, Angels, and Victorians: The Study of Darwin, Huxley, and Evolution*. New York: McGraw-Hill.

Iversen, Erik (1993), *The Myth of Egypt and its Hieroglyphs in European Tradition*. Princeton, NJ: Princeton University Press.

Jacks, Philip J. (1993), *The Antiquarian and the Myth of Antiquity: The Origins of Rome in Renaissance Thought*. Cambridge: Cambridge University Press.

Jacquard, Albert (1985), *Endangered by Science?* New York: Columbia University Press.

Johnson, Gregory A. (1973), *Local Exchange and Early State Development in Southwestern Iran*. Ann Arbor, MI: University of Michigan, Museum of Anthropology, Anthropological Papers, 51.

Johnson, Matthew (1996), *An Archaeology of Capitalism*. Oxford: Basil Blackwell.

Karlen, Arno (1995), *Man and Microbes: Diseases and Plagues in History and Modern Times*. New York: Putnam.

Keating, Michael (1986), *To the Last Drop: Canada and the World's Water Crisis*. Toronto: Macmillan.

Kedourie, Elie (1960), *Nationalism*. London: Hutchinson.

Keen, Benjamin (1971), *The Aztec Image in Western Thought*. New Brunswick, NJ: Rutgers University Press.

Kemble, John M. (1849), *The Saxons in England: A History of the English Commonwealth till the Period of the Norman Conquest*, 2 vols. London: Longman, Brown, Green and Longmans.

Kendrick, Thomas D. (1950), *British Antiquity*. London: Methuen.

Kidder, Alfred V. (1924), *An Introduction to the Study of Southwestern Archaeology*. New Haven: Papers of the Southwestern Expedition, Phillips Academy, 1 (reprinted 1962 with an Introduction 'Southwestern archaeology today', by Irving Rouse. New Haven, CT: Yale University Press).

—— (1940), Looking backward. *Proceedings of the American Philosophical Society*, 83, pp. 527–37.

Kiser, Edgar, and Michael Hechter (1991), The role of general theory in comparative-historical sociology. *American Journal of Sociology*, 97, pp. 1–30.

280 REFERENCES

Klemm, Gustav F. (1843–52), *Allgemeine Cultur-Geschichte der Mensch-heit*, 10 vols. Leipzig: Teubner.

Klindt-Jensen, Ole (1975), *A History of Scandinavian Archaeology*. London: Thames and Hudson.

Kluckhohn, Clyde (1960), The moral order in the expanding society. In Carl H. Kraeling and Robert M. Adams (eds), *City Invincible*, pp. 391–404. Chicago, IL: University of Chicago Press.

Knorr-Cetina, Karin D. (1981), *The Manufacture of Knowledge: An Essay on the Constructivist and Contextual Nature of Science*. Oxford: Pergamon.

Knox, Robert (1862), *The Races of Men: A Philosophical Enquiry into the Influence of Race over the Destinies of Nations*, 2nd edn. London: Renshaw (1st edn 1850).

Kossinna, Gustaf (1911), *Die Herkunft der Germanen*. Leipzig: Kabitzsch.

Krieger, Joel (1986), *Reagan, Thatcher, and the Politics of Decline*. Cambridge: Polity Press.

Kroeber, Alfred L. (1948), *Anthropology*, 2nd edn. New York: Harcourt, Brace and Company.

—— (1952), *The Nature of Culture*. Chicago, IL: University of Chicago Press.

Kroker, Arthur (1984), *Technology and the Canadian Mind: Innis/McLuhan/Grant*. Montreal: New World Perspectives.

Kropotkin, Petr A. (1902), *Mutual Aid: A Factor of Evolution*. New York: McClure, Phillips.

Lafitau, Joseph-François (1724), *Moeurs des sauvages amériquains comparées aux moeurs des premiers temps*. Paris: Saugrain l'aîné.

Laming-Emperaire, Annette (1964), *Origines de l'archéologie préhistorique en France des superstitions médiévales à la découverte de l'homme fossile*. Paris: Picard.

Latour, Bruno, and Steve Woolgar (1979), *Laboratory Life: The Social Construction of Scientific Facts*. Beverly Hills, CA: Sage.

Laudan, Larry (1990), *Science and Relativism: Some Key Controversies in the Philosophy of Science*. Chicago, IL: University of Chicago Press.

Laughlin, Charles D., Jr., and Ivan A. Brady (eds) (1978), *Extinction and Survival in Human Populations*. New York: Columbia University Press.

Leach, Edmund (1970), *Lévi-Strauss*. London: Collins.

Lee, Richard B. (1990), Primitive communism and the origin of social inequality. In Steadman Upham (ed.), *The Evolution of Political Sys-*

tems: Sociopolitics in Small-Scale Sedentary Societies, pp. 225–46. Cambridge: Cambridge University Press.

Lee, Richard B., and Irven DeVore (eds) (1968), *Man the Hunter*. Chicago, IL: Aldine.

Leiss, William (1990), *Under Technology's Thumb*. Montreal: McGill-Queen's University Press.

Lemann, Nicholas (1997), Citizen 501(c)(3): An increasingly powerful agent in American life is also one of the least noticed. *Atlantic Monthly*, 279(2), pp. 18–20.

Lenin, Vladimir I. [1917] (1939), *Imperialism, The Highest Stage of Capitalism: A Popular Outline*. New York: International Publishers.

Lescarbot, Marc [1617–18] (1907–14), *The History of New France*, tr. William L. Grant, 3 vols. Toronto: Champlain Society.

Lévi-Strauss, Claude (1985), *The View from Afar*. Oxford: Basil Blackwell.

Levitt, Kari, and Alister McIntyre (1967), *Canada–West Indies Economic Relations*. Montreal: Centre for Developing-Area Studies, McGill University.

Lévy-Bruhl, Lucien (1922), *La Mentalité primitive*. Paris: Alcan.

Locke, John [1690] (1952), *The Second Treatise of Government*, ed. Thomas E. Peardon. New York: Liberal Arts Press.

Lowie, Robert H. (1920), *Primitive Society*. New York: Bonie and Liveright.

—— (1959), *Robert H. Lowie, Ethnologist: A Personal Record*. Berkeley, CA: University of California Press.

Lubbock, John (1869), *Pre-historic Times, as Illustrated by Ancient Remains, and the Manners and Customs of Modern Savages*, 2nd edn. London: Williams and Norgate.

—— (1870), *The Origin of Civilisation and the Primitive Condition of Man*. London: Longmans, Green.

Luttwak, Edward (1996), Central bankism. *London Review of Books*, 18(22), pp. 3–7.

Lyell, Charles (1863), *The Geological Evidences of the Antiquity of Man, with Remarks on Theories of the Origin of Species by Variation*. London: John Murray.

Macaulay, Thomas B. (1858), *The History of England from the Accession of James the Second*, 7 vols. London: Longman, Brown, Green, Longmans and Roberts.

MacGaffey, Wyatt (1966), Concepts of race in the historiography of northeast Africa. *Journal of African History*, 7, pp. 1–17.

McGuire, Randall H. (1983), Breaking down cultural complexity: Inequality and heterogeneity. In Michael B. Schiffer (ed.), *Advances in Archaeological Method and Theory*, 6, pp. 91–142. New York: Academic Press.

McLaren, Angus (1986), The creation of a haven for 'human thoroughbreds': The sterilization of the feeble-minded and the mentally ill in British Columbia. *Canadian Historical Review*, 67, pp. 127–50.

McLennan, John P. (1865), *Primitive Marriage: An Inquiry into the Origin of the Form of Capture in Marriage Ceremonies*. Edinburgh: Adam and Charles Black.

McLuhan, Marshall (1951), *The Mechanical Bride: Folklore of Industrial Man*. New York: Vanguard Press.

—— (1962), *The Gutenberg Galaxy: The Making of Typographic Man*. Toronto: University of Toronto Press.

McMichael, Anthony J. (1995), *Planetary Overload: Global Environmental Change and the Health of the Human Species*. Cambridge: Cambridge University Press.

McNeill, William H. (1976), *Plagues and Peoples*. Garden City, NY: Anchor Press.

—— (1986), *Mythistory and Other Essays*. Chicago, IL: University of Chicago Press.

MacNeish, Richard S. (1978), *The Science of Archaeology?* North Scituate, MA: Duxbury Press.

Maine, Henry J. S. (1861), *Ancient Law: Its Connection with the Early History of Society, and its Relation to Modern Ideas*. London: John Murray.

Mandelbaum, Maurice (1971), History, Man, and Reason: A Study in Nineteenth-Century Thought. Baltimore, MD: Johns Hopkins University Press.

Mann, Michael (1986), *The Sources of Social Power*, vol. 1, *A History of Power from the Beginning to A.D. 1760*. Cambridge: Cambridge University Press.

Marchak, M. Patricia (1991), *The Integrated Circus: The New Right and the Restructuring of Global Markets*. Montreal: McGill-Queen's University Press.

—— (1995), *Logging the Globe*. Montreal: McGill-Queen's University Press.

Marcuse, Herbert (1964), *One Dimensional Man: Studies in the Ideology of Advanced Industrial Society*. Boston, MA: Beacon Press.

Marsden, Barry (1974), *The Early Barrow-Diggers*. Park Ridge, NY: Noyes Press.

Martin, Paul S., and Richard G. Klein (eds) (1982), *Quaternary Extinctions: A Prehistoric Revolution*. Tucson, AZ: University of Arizona Press.

Marx, Karl, and Frederick Engels (1962), *Selected Works in Two Volumes*. Moscow: Foreign Languages Publishing House.

Maschner, Herbert D. (ed.) (1996), *Darwinian Archaeologies*. New York: Plenum.

Meek, Ronald L. (1976), *Social Science and the Ignoble Savage*. Cambridge: Cambridge University Press.

Melleuish, Gregory (1995), The place of Vere Gordon Childe in Australian intellectual history. In Peter Gathercole, Terry H. Irving, and Gregory Melleuish (eds), *Childe and Australia: Archaeology, Politics and Ideas*, pp. 147–61. St Lucia, Australia: University of Queensland Press.

Merrell, James H. (1989), *The Indians' New World: Catawbas and their Neighbors from European Contact through the Era of Removal*. Chapel Hill, NC: University of North Carolina Press.

Mintz, Sidney W. (1985), *History, Evolution, and the Concept of Culture: Selected Papers by Alexander Lesser*. Cambridge: Cambridge University Press.

Montaigne, Michel de (1965), *The Complete Essays of Montaigne*, tr. Donald M. Frame. Stanford, CA: Stanford University Press.

Montesquieu, Charles-Louis de Secondat, Baron de (1748), *De l'esprit des lois*. Geneva: Barrillot.

Moran, Paul, and David S. Hides (1990), Writing, authority and the determination of a subject. In Ian Bapty and Tim Yates (eds), *Archaeology after Structuralism: Post-Structuralism and the Practice of Archaeology*, pp. 205–20. London: Routledge.

Morgan, Lewis Henry (1877), *Ancient Society; Or, Researches in the Lines of Human Progress from Savagery through Barbarism to Civilization*. New York: Holt.

Moseley, Michael E. (1975), *The Maritime Foundations of Andean Civilization*. Menlo Park, CA: Cummings.

Movius, Hallam L., Jr. (1948), The Lower Palaeolithic cultures of southern and eastern Asia. *Transactions of the American Philosophical Society*, 38, pp. 329–420.

Murdock, George P. (1945), The common denominator of cultures. In

Ralph Linton (ed.), *The Science of Man in the World Crisis*, pp. 123–42. New York: Columbia University Press.

—— (1949), *Social Structure*. New York: Macmillan.

—— (1956), How culture changes. In Harry L. Shapiro (ed.), *Man, Culture, and Society*, pp. 247–60. New York: Oxford University Press.

—— (1959), Evolution in social organization. In Betty Meggers (ed.), *Evolution and Anthropology: A Centennial Appraisal*, pp. 126–43. Washington, DC: Anthropological Society of Washington.

Myers, Norman (1984), *Gaia: An Atlas of Planet Management*. Garden City, NY: Doubleday.

Myres, John L. (1911), *The Dawn of History*. London: Williams and Norgate.

Nash, Ronald J. (1995), Deconstructing archaeology. *Canadian Journal of Archaeology*, 19, pp. 19–28.

Nilsson, Sven (1868), *The Primitive Inhabitants of Scandinavia: An Essay on Comparative Ethnography*, 3rd edn. London: Longmans, Green.

Nisbet, Robert A. (1969), *Social Change and History: Aspects of the Western Theory of Development*. Oxford: Oxford University Press.

Nitecki, Matthew H. (ed.) (1984), *Extinctions*. Chicago, IL: University of Chicago Press.

North, Richard (1995), *Life on a Modern Planet: A Manifesto for Progress*. Manchester: Manchester University Press.

O'Brien, Michael J. (1996), *Evolutionary Archaeology: Theory and Application*. Salt Lake City, UT: University of Utah Press.

O'Hara, Sarah L., F. Alayne Street-Perrot, and Timothy P. Burt (1993), Accelerated soil erosion around a Mexican highland lake caused by prehispanic agriculture. *Nature*, 362, pp. 48–51.

Pagden, Anthony (1982), *The Fall of Natural Man: The American Indian and the Origins of Comparative Ethnology*. Cambridge: Cambridge University Press.

Parsons, Jeffrey R., Elizabeth Brumfiel, Mary H. Parsons, and David J. Wilson (1982), *Prehispanic Settlement Patterns in the Southern Valley of Mexico: The Chalco–Xochimilco Region*. Ann Arbor, MI: University of Michigan, Memoirs of the Museum of Anthropology, 14.

Parsons, Talcott (1966), *Societies: Evolutionary and Comparative Perspectives*. Englewood Cliffs, NJ: Prentice-Hall.

—— (1971), *The System of Modern Societies*. Englewood Cliffs, NJ: Prentice-Hall.

Peace, William J. (1988), Vere Gordon Childe and American anthropology. *Journal of Anthropological Research*, 44, pp. 417–33.

Peake, Harold J. E., and Herbert J. Fleure (1927), *The Corridors of Time*, vol. 3, *Peasants and Potters*. Oxford: Oxford University Press.

Peel, John D. (1971), *Herbert Spencer: The Evolution of a Sociologist*. London: Heinemann Educational.

Pepper, David (1993), *Eco-Socialism: From Deep Ecology to Social Justice*. London: Routledge.

Perry, William J. (1923), *The Children of the Sun: A Study in the Early History of Civilization*. London: Methuen.

—— (1924), *The Growth of Civilization*. London: Methuen.

Petrie, W. M. Flinders (1939), *The Making of Egypt*. London: Sheldon.

Piggott, Stuart (1985), *William Stukeley: An Eighteenth-Century Antiquary*, 2nd edn. London: Thames and Hudson.

Pitt-Rivers, Augustus H. L.-F. (1858), On the improvement of the rifle as a weapon for general use. *Journal of the Royal United Services Institution*, 2, pp. 453–88.

—— (1906), *The Evolution of Culture and Other Essays*. Oxford: Oxford University Press.

Polanyi, Karl (1944), *The Great Transformation*. New York: Farrar and Rinehart.

Polanyi, Karl, Conrad M. Arensberg, and Harry W. Pearson (eds) (1957), *Trade and Market in the Early Empires: Economics in History and Theory*. Glencoe, IL: Free Press.

Polgar, Steven (ed.) (1975), *Population, Ecology, and Social Evolution*. The Hague: Mouton.

Popper, Karl R. (1957), *The Poverty of Historicism*. Boston, MA: Beacon Press.

Potter, Rob, and Barbara Welch (1996), Caribbean indigenization and development. *Caribbean Week*, 8(01), pp. 13–14.

Pumpelly, Raphael (ed.) (1908), *Explorations in Turkestan: Expedition of 1904*, 2 vols. Washington, DC: Carnegie Institution.

Raglan, Fitzroy R., Baron (1939), *How Came Civilization?* London: Methuen.

Rappaport, Roy A. (1968), *Pigs for the Ancestors: Ritual in the Ecology of a New Guinea People*. New Haven, CT: Yale University Press.

—— (1979), *Ecology, Meaning, and Religion*. Richmond, CA: North Atlantic Books.

Rathje, William L. (1975), The last tango in Mayapán: A tentative trajectory of production-distribution systems. In Jeremy A. Sabloff and C. C. Lamberg-Karlovsky (eds), *Ancient Civilization and Trade*, pp. 409–48. Albuquerque, NM: University of New Mexico Press.

Ray, Arthur J. (1974), *Indians in the Fur Trade: Their Role as Trappers, Hunters, and Middlemen in the Lands Southwest of Hudson Bay, 1660–1870*. Toronto: University of Toronto Press.

Reeves, Marjorie (1969), *The Influence of Prophecy in the Later Middle Ages: A Study in Joachimism*. Oxford: Oxford University Press.

Renfrew, A. Colin (1973), *Before Civilization: The Radiocarbon Revolution and Prehistoric Europe*. London: Cape.

—— (1978), Trajectory discontinuity and morphogenesis. *American Antiquity*, 43, pp. 203–22.

—— (1988), Towards a world prehistory. In Glyn Daniel and Colin Renfrew, *The Idea of Prehistory*, 2nd edn., pp. 176–204. Edinburgh: University of Edinburgh Press.

Renfrew, A. Colin, and John F. Cherry (eds) (1986), *Peer Polity Interaction and Socio-political Change*. Cambridge: Cambridge University Press.

Renfrew, A. Colin, and Stephen Shennan (eds) (1982), *Ranking, Resource and Exchange: Aspects of the Archaeology of Early European Society*. Cambridge: Cambridge University Press.

Richardson, Boyce (1990), *Time to Change: Canada's Place in a World in Crisis*. Toronto: Summerhill Press.

Ridley, Matt (1995), *Down to Earth: A Contrarian View of Environmental Problems*. London: Institute of Economic Affairs.

Rindos, David (1984), *The Origins of Agriculture: An Evolutionary Perspective*. New York: Academic Press.

Roberts, Clayton (1996), *The Logic of Historical Explanation*. University Park, PA: Pennsylvania State University Press.

Rose, Margaret A. (1991), *The Post-Modern and the Post-Industrial: A Critical Analysis*. Cambridge: Cambridge University Press.

Rosenau, Pauline M. (1992), *Post-Modernism and the Social Sciences: Insights, Inroads, and Intrusions*. Princeton, NJ: Princeton University Press.

Rosenthal, Franz (ed.) (1967), *Ibn Khaldûn: The Muqaddimah: An Introduction to History*. London: Routledge and Kegan Paul.

Rossi, Paolo (1984), *The Dark Abyss of Time: The History of the Earth and the History of Nations from Hooke to Vico*. Chicago, IL: University of Chicago Press.

Rouse, Irving B. (1972), *Introduction to Prehistory: A Systematic Approach*. New York: McGraw-Hill.

Rowell, Andrew (1996), *Green Backlash: Global Subversion of the Environment Movement*. London: Routledge.

Rowlands, Michael J. (1989), A question of complexity. In Daniel Miller, Michael Rowlands, and Christopher Tilley (eds), *Domination and Resistance*, pp. 29–40. London: Unwin Hyman.

Rowlands, Michael, Mogens Larsen, and Kristian Kristiansen (eds) (1987), *Centre and Periphery in the Ancient World*. Cambridge: Cambridge University Press.

Sabloff, Jeremy A., and C. C. Lamberg-Karlovsky (eds) (1975), *Ancient Civilization and Trade*. Albuquerque, NM: University of New Mexico Press.

Sahlins, Marshall D. (1958), *Social Stratification in Polynesia*. Seattle, WA: University of Washington Press.

—— (1968), *Tribesmen*. Englewood Cliffs, NJ: Prentice-Hall.

—— (1972), *Stone Age Economics*. Chicago, IL: Aldine.

—— (1976), *Culture and Practical Reason*. Chicago, IL: University of Chicago Press.

Sahlins, Marshall D., and Elman R. Service (eds) (1960), *Evolution and Culture*. Ann Arbor, MI: University of Michigan Press.

Said, Edward W. (1994), *Representations of the Intellectual*. New York: Pantheon Books.

Salisbury, Richard F. (1986), *A Homeland for the Cree: Regional Development in James Bay, 1971–1981*. Montreal: McGill-Queen's University Press.

Sanders, Scott R. (1995), *Writing from the Center*. Bloomington, IN: Indiana University Press.

Sanders, William T., Jeffrey R. Parsons, and Robert S. Santley (1979), *The Basin of Mexico: Ecological Processes in the Evolution of a Civilization*. New York: Academic Press.

Sanderson, Stephen K. (1990), *Social Evolutionism: A Critical History*. Oxford: Basil Blackwell.

Sassen, Saskia (1996), *Losing Control? Sovereignty in an Age of Globalization*. New York: Columbia University Press.

Saul, John R. (1995), *The Unconscious Civilization*. Concord, ON: Anansi Press.

Schnapp, Alain (1993), *La Conquête du passé: Aux origines de l'archéologie*. Paris: Editions Carré.

Schrire, Carmel (ed.) (1984), *Past and Present in Hunter Gatherer Studies*. Orlando, FL: Academic Press.

Sebag, Lucien (1964), *Marxisme et structuralisme*. Paris: Payot.

Service, Elman R. (1962), *Primitive Social Organization: An Evolutionary Perspective*. New York: Random House.

288   REFERENCES

—— (1975), *Origins of the State and Civilization: The Process of Cultural Evolution*. New York: Norton.

—— (1985), *A Century of Controversy: Ethnological Issues from 1860 to 1960*. Orlando, FL: Academic Press.

Seton, Ernest T. (1936), *The Gospel of the Red Man: An Indian Bible*. New York: Doubleday.

Shanks, Michael, and Christopher Tilley (1987), *Social Theory and Archaeology*. Cambridge: Polity Press.

Shapiro, Harry L. (1952), Revised version of UNESCO statement on race. *American Journal of Physical Anthropology*, 10, pp. 363–8.

Shennan, Stephen (1993), After social evolution: The new archaeological agenda? In Norman Yoffee and Andrew Sherratt (eds), *Archaeological Theory: Who Sets the Agenda?*, pp. 53–9. Cambridge: Cambridge University Press.

Shepherd, Linda J. (1993), *Lifting the Veil: The Feminine Face of Science*. Boston: Shambhala.

Sherratt, Andrew (1993), The relativity of theory. In Norman Yoffee and Andrew Sherratt (eds), *Archaeological Theory: Who Sets the Agenda?*, pp. 119–30. Cambridge: Cambridge University Press.

Shnirelman, Victor A. (1995), From internationalism to nationalism: Forgotten pages of Soviet archaeology in the 1930s and 1940s. In Philip L. Kohl and Clare Fawcett (eds), *Nationalism, Politics, and the Practice of Archaeology*, pp. 120–38. Cambridge: Cambridge University Press.

Silverberg, Robert (1968), *Mound Builders of Ancient America: The Archaeology of a Myth*. Greenwich, CT: New York Graphic Society.

Simmons, Ian G. (1993), *Environmental History: A Concise Introduction*. Oxford: Basil Blackwell.

Sioui, Georges E. (1992), *For an Amerindian Autohistory: An Essay on the Foundations of a Social Ethic*. Montreal: McGill-Queen's University Press.

Slotkin, James S. (1965), *Readings in Early Anthropology*. New York: Viking Fund Publications in Anthropology, 40.

Smith, Anthony D. (1973), *The Concept of Social Change: A Critique of the Functionalist Theory of Social Change*. London: Routledge and Kegan Paul.

Smith, Grafton Elliot (1923), *The Ancient Egyptians and the Origin of Civilization*. London: Harper.

Smith, Susan J. (1986), Police accountability and local democracy. *Area*, 18, pp. 99–107.

Sollas, William J. (1911), *Ancient Hunters and their Modern Representatives*. London: Macmillan.

Solway, David (1997), *Lying about the Wolf: Essays in Culture and Education*. Montreal: McGill-Queen's University Press.

Soros, George (1997), The capitalist threat. *Atlantic Monthly*, 279(2), pp. 45–58.

Speer, Albert (1971), *Inside the Third Reich*. New York: Avon Books.

Spengler, Oswald [1918–22] (1926–8), *The Decline of the West*, 2 vols. New York: Alfred Knopf.

Sperber, Dan (1985), Anthropology and psychology: Towards an epidemiology of representations. *Man*, 20, pp. 73–89.

Spinden, Herbert J. (1928), *Ancient Civilizations of Mexico and Central America*. New York: American Museum of Natural History, Handbook Series, 3.

Stanton, William R. (1960), *The Leopard's Spots: Scientific Attitudes toward Race in America, 1815–59*. Chicago, IL: University of Chicago Press.

Steadman, Philip (1979), *The Evolution of Designs: Biological Analogy in Architecture and the Applied Arts*. Cambridge: Cambridge University Press.

Stein, George J. (1988), Biological science and the roots of Nazism. *American Scientist*, 76, pp. 50–8.

Stevens, William (1993), The Eden myth. *The Gazette* (Montreal), 3 April.

Steward, Julian H. (1949), Cultural causality and law: A trial formulation of the development of early civilizations. *American Anthropologist*, 51, pp. 1–27.

—— (1955), *Theory of Culture Change: The Methodology of Multilinear Evolution*. Urbana, IL: University of Illinois Press.

—— (1960), Some implications of the symposium. In Julian H. Steward *et al.* (eds), *Irrigation Civilizations: A Comparative Study*, pp. 58–78. Washington, DC: Pan American Union, Social Science Monographs, 1.

Stocking, George W., Jr. (1968), *Race, Culture, and Evolution: Essays in the History of Anthropology*. New York: Free Press.

—— (1987), *Victorian Anthropology*. New York: Free Press.

Stubbs, William (1891–1903), *The Constitutional History of England in its Origin and Development*, 5th edn, 3 vols. Oxford: Clarendon Press.

Sumner, William G. (1906), *Folkways*. Boston: Ginn.

Tainter, Joseph A. (1988), *The Collapse of Complex Societies*. Cambridge: Cambridge University Press.

Tanner, Adrian (1979), *Bringing Home Animals: Religious Ideology and Mode of Production of the Mistassini Cree Hunters*. London: Hurst.

Taylor, Bron R. (ed.) (1995), *Ecological Resistance Movements: The Global Emergence of Radical and Popular Environmentalism*. Albany, NY: State University of New York Press.

Teilhard de Chardin, Pierre [1938–40] (1959), *The Phenomenon of Man*. New York: Harper.

Teltser, Patrice A. (ed.) (1995), *Evolutionary Archaeology: Methodological Issues*. Tucson, AZ: University of Arizona Press.

Tenner, Edward (1996), *Why Things Bite Back: Technology and the Revenge of Unintended Consequences*. New York: Knopf.

Terrell, John (1988), History as a family tree, history as a tangled bank: Constructing images and interpretations of prehistory in the South Pacific. *Antiquity*, 62, pp. 642–57.

Thomas, David H. (1972), A computer simulation model of Great Basin Shoshonean subsistence and settlement patterns. In David L. Clarke (ed.), *Models in Archaeology*, pp. 671–704. London: Methuen.

Thwaites, Reuben G. (1896–1901), *The Jesuit Relations and Allied Documents*, 73 vols. Cleveland, OH: Burrows Brothers.

Tice, D. J. (1996), Ties that bind no more. *The Gazette* (Montreal), 13 January, p. B-6.

Tilley, Christopher (1995), Clowns and circus acts. *Critique of Anthropology*, 15, pp. 337–41.

Toffler, Alvin (1970), *Future Shock*. New York: Random House.

Toulmin, Stephen E., and June Goodfield (1966), *The Discovery of Time*. New York: Harper and Row.

Toynbee, Arnold J. (1934–61), *A Study of History*, 12 vols. Oxford: Oxford University Press.

Trigger, Bruce G. (1980), Archaeology and the image of the American Indian. *American Antiquity*, 45, pp. 662–76.

—— (1985), *Natives and Newcomers: Canada's 'Heroic Age' Reconsidered*. Montreal: McGill-Queen's University Press.

—— (1989), *A History of Archaeological Thought*. Cambridge: Cambridge University Press.

—— (1990), Maintaining economic equality in opposition to complexity: An Iroquoian case study. In Steadman Upham (ed.), *The Evolution of Political Systems: Sociopolitics in Small-Scale Sedentary Societies*, pp. 119–45. Cambridge: Cambridge University Press.

—— (1993), *Early Civilizations: Ancient Egypt in Context*. Cairo: American University in Cairo Press.

Turner, Sharon [1799–1805] (1823), *The History of the Anglo-Saxons: Comprising the History of England from the Earliest Period to the Norman Conquest*, 4th edn, 3 vols. London: Longman, Hurst, Rees, Orme, and Browne.

Tylor, Edward B. (1889), On a method of investigating the development of institutions; applied to laws of marriage and descent. *Journal of the Anthropological Institute of Great Britain and Ireland*, 18, pp. 245–72.

Ucko, Peter J., Michael Hunter, Alan J. Clark, and Andrew David (1991), *Avebury Reconsidered: From the 1660s to the 1990s*. London: Unwin Hyman.

Ucko, Peter J., and Andrée Rosenfeld (1967), *Palaeolithic Cave Art*. London: Weidenfeld and Nicolson.

UNESCO Statement on Race. *See* Shapiro (1952).

Van Riper, A. Bowdoin (1993), *Men among the Mammoths: Victorian Science and the Discovery of Human Prehistory*. Chicago, IL: University of Chicago Press.

Vaughan, Alden T. (1982), From white man to red skin: Changing Anglo-American perceptions of the American Indian. *American Historical Review*, 87, pp. 917–53.

Vavilov, Nicholai I. (1951), *The Origin, Variation, Immunity, and Breeding of Cultivated Plants: Selected Writings*. Waltham, MA: Chronica Botanica Company.

von Däniken, Erich (1970), *Chariots of the Gods?: Unsolved Mysteries of the Past*. New York: Putnam.

Wallace, Anthony F. C. (1961), *Culture and Personality*. New York: Random House.

Wallerstein, Immanuel (1974), *The Modern World-System*, vol. 1, *Capitalist Agriculture and the Origins of the European World-Economy in the Sixteenth Century*. New York: Academic Press.

Watson, Patty Jo, Steven A. LeBlanc, and Charles L. Redman (1971), *Explanation in Archeology: An Explicitly Scientific Approach*. New York: Columbia University Press.

Weiss, Roberto (1969), *The Renaissance Discovery of Classical Antiquity*. Oxford: Basil Blackwell.

Welch, Barbara M. (1996a), Table d'hôte or à la carte? French and Commonwealth approaches to development in the central Lesser Antilles. *Canadian Geographer*, 40, pp. 136–47.

—— (1996b), *Survival by Association: Supply Management Landscapes of the Eastern Caribbean*. Montreal: McGill-Queen's University Press.

Wenke, Robert J. (1981), Explaining the evolution of cultural complexity: A review. In Michael B. Schiffer (ed.), *Advances in Archaeological Method and Theory*, 4, pp. 79–127. New York: Academic Press.

White, Leslie A. (1945), 'Diffusion vs evolution': An anti-evolutionist fallacy. *American Anthropologist*, 47, pp. 339–51.

—— (1949), *The Science of Culture: A Study of Man and Civilization*. New York: Farrar, Straus.

—— (1959), *The Evolution of Culture: The Development of Civilization to the Fall of Rome*. New York: McGraw-Hill.

—— (1975), *The Concept of Cultural Systems: A Key to Understanding Tribes and Nations*. New York: Columbia University Press.

White, Lynn, Jr. (1962), *Medieval Technology and Social Change*. Oxford: Oxford University Press.

—— (1967), The historical roots of our ecological crisis. *Science*, 155, pp. 1203–7.

Whitehead, Neil L. (1995), An interview with Jan Vansina. *Ethnohistory*, 42, pp. 303–16.

Widmer, Randolph J. (1988), *The Evolution of the Calusa: A Non-agricultural Chiefdom on the Southwest Florida Coast*. Tuscaloosa, AL: University of Alabama Press.

Wilson, Daniel (1862), *Prehistoric Man: Researches into the Origin of Civilisation in the Old and the New World*. Cambridge: Macmillan.

Wilson, Edward O. (1978), *On Human Nature*. Cambridge, MA: Harvard University Press.

Winckelmann, Johann J. [1764] (1968), *History of Ancient Art*, 2 vols. New York: Ungar.

Wittfogel, Karl (1957), *Oriental Despotism: A Comparative Study of Total Power*. New Haven, CT: Yale University Press.

Wolf, Eric R. (1982), *Europe and the People without History*. Berkeley, CA: University of California Press.

Woolfson, Charles (1982), *The Labour Theory of Culture: A Re-examination of Engels's Theory of Human Origins*. London: Routledge and Kegan Paul.

Wrangham, Richard W., and Dale Peterson (1996), *Dominant Males: Apes and the Origins of Human Violence*. Boston, MA: Houghton Mifflin.

Wylie, Alison (1982), Epistemological issues raised by a structuralist archaeology. In Ian Hodder (ed.), *Symbolic and Structural Archaeology*, pp. 39–46. Cambridge: Cambridge University Press.

—— (1985), Facts of the record and facts of the past: Mandelbaum on the anatomy of history 'proper'. *International Studies in Philosophy*, 17, pp. 71–85.

—— (1993), A proliferation of new archaeologies: 'Beyond objectivism and relativism'. In Norman Yoffee and Andrew Sherratt (eds), *Archaeological Theory: Who Sets the Agenda?*, pp. 20–6. Cambridge: Cambridge University Press.

Yates, Frances A. (1964), *Giordano Bruno and the Hermetic Tradition*. London: Routledge and Kegan Paul.

Yoffee, Norman (1988), Orienting collapse. In Norman Yoffee and George L. Cowgill (eds), *The Collapse of Ancient States and Civilizations*, pp. 1–19. Tucson, AZ: University of Arizona Press.

—— (1993), Too many chiefs? (Or, safe texts for the '90s). In Norman Yoffee and Andrew Sherratt (eds), *Archaeological Theory: Who Sets the Agenda?*, pp. 60–78. Cambridge: Cambridge University Press.

Yoffee, Norman, and George L. Cowgill (eds) (1988), *The Collapse of Ancient States and Civilizations*. Tucson, AZ: University of Arizona Press.

Yoffee, Norman, and Andrew Sherratt (eds) (1993), *Archaeological Theory: Who Sets the Agenda?* Cambridge: Cambridge University Press.

Zimmerman, Michael E. (1994), *Contesting Earth's Future: Radical Ecology and Postmodernity*. Berkeley, CA: University of California Press.

Index

aboriginal peoples, *see* indigenous peoples, Indians
absolutism, 3, 31, 34–5, 53, 187
acculturation, 64, 73, 95, 163, 172, 175, 208, 235–6, 261
Acosta, J. de, 22, 28
Act of Union (1707), 31–2
Adams, R. N., 147–8, 199
Adams, R. McC., 146
adaptation, ecological: biological, 9, 32–3, 60, 71–2, 114, 136, 200; cultural, 13, 57, 77, 111, 114, 121–2, 126, 128, 130, 133, 135–6, 139, 155–6, 158, 161, 168–9, 177, 184, 209, 227–8, 231–3, 240–1, 249, 254, 260
agriculture: 18, 22, 33, 37, 76, 80, 100–1, 110–11, 129, 140, 143–4, 146, 163–4, 171, 180, 192, 195–7, 220, 231–3, 236; agribusiness, 196
Aids, 189
alienation, 44, 87, 92, 190–1, 212
altruism, 27, 62, 183–4, 216, 235, 257
American Revolution, 42, 235
Americas, *see* New World
analogy, analogue, 160–3, 224–5, 250; *see also* organic analogy
Anaximander of Miletus, 2
anthropocentrism, 33, 170, 213–14, 231
anthropology, anthropologist, 11–12, 15, 45, 77–8, 80, 86, 97–102, 109, 113, 124, 126, 129, 134, 145, 154, 162–3, 177, 211, 248, 251; *see also* archaeology,

cultural history, ethnology, physical anthropology
anti-ecologism, 230–3
antitrust laws, 194
arboriculture, 196–7
archaeology, archaeologists, 26, 28, 36–7, 40, 51–3, 59, 71, 73–4, 83, 94–7, 102–8, 109–13, 122, 124, 131, 133–4, 136, 140, 145–7, 163, 186, 228, 251
archetypes, *see* essentialism
architecture, 19–21, 163–4
Ardrey, R., 182
Argyll, Duke of, 62
Arouet, *see* Voltaire
art, artists, 19–21, 46, 74, 184
Aryans, 106
'Asian culture', as a modern political construct, 239
Asiatic mode of production, 92
atomic energy, *see* energy
Aztecs, 180–1

Bachofen, J., 74
Bacon, F., 25, 43, 121
Bagehot, W., 85
band societies, 130, 160, 211, 238
basic law of evolution (L. White), 127
Bastian, A., 78, 97
behaviourism, 98
Benedict, R., 98
Bentham, J., 61, 90
Bible, biblical traditions, 16–19, 21–3, 28, 37, 51, 59, 225, 228, 231